Dark Count Rate of Silicon Phot.

UNIVERSITÄT DER BUNDESWEHR MÜNCHEN
Fakultät für Elektrotechnik und Informationstechnik
Institut für Physik

DARK COUNT RATE OF SILICON PHOTOMULTIPLIERS

- Metrological Characterization and Suppression -

Eugen Engelmann

Vollständiger Abdruck der von der Fakultät für
Eletrotechnik und Informationstechnik, Institut für Physik
der Universität der Bundeswehr München
zur Erlangung des akademischen Grades eines

Doktor der Naturwissenschaften
(Dr. rer. nat.)

genehmigten Dissertation.

Vorsitzender: Prof. Dr. Linus Maurer
1. Gutachter: Prof. Dr. Walter Hansch
2. Gutachter: Prof. Dr. Erika Garutti

Die Dissertation wurde am 17.01.2018 bei der
Universität der Bundeswehr München eingereicht
und durch die Fakultät für Elektrotechnik und Informationstechnik
am 05.06.2018 angenommen. Die mündliche Prüfung fand am 02.07.2018 statt.

Bibliografische Information der Deutschen Nationalbibliothek

Die Deutsche Nationalbibliothek verzeichnet diese Publikation in der Deutschen
Nationalbibliografie; detaillierte bibliografische Daten sind im Internet über
http://dnb.d-nb.de abrufbar.

1. Aufl. - Göttingen: Cuvillier, 2018
Zugl.: München, Univ. der Bundeswehr, Diss., 2018

© CUVILLIER VERLAG, Göttingen 2018
   Nonnenstieg 8, 37075 Göttingen
   Telefon: 0551-54724-0
   Telefax: 0551-54724-21
   www.cuvillier.de

1. Auflage, 2018
Gedruckt auf umweltfreundlichem, säurefreiem Papier aus nachhaltiger Forstwirtschaft

   ISBN 978-3-7369-9892-6
   eISBN 978-3-7369-8892-7

Für meine Familie

*[...] Die Wissenschaft dagegen nötigt uns,
den Glauben an einfache Kausalitäten
gerade dort aufzugeben, wo alles so leicht
begreiflich scheint und wir die Narren
des Augenscheins sind [...]*

F. Nietzsche, "Morgenröte"

Abstract

In recent years, Silicon Photomultipliers (SiPMs) developed into mature photon detectors. Their high photon detection efficiency, insensitivity to magnetic fields, fast timing properties, compactness and robustness make them the detector of choice in a wide field of applications. However, the noise of SiPMs, which is determined by the dark count rate, the optical crosstalk (prompt and delayed) and the afterpulsing, sets limitations to detector systems which use SiPMs.

The main subject of this work is the dark count rate of Silicon Photomultipliers. This quantity describes the rate at which pulses are generated in the absence of light and deteriorate the photon counting performance of SiPMs. To identify and evaluate the individual physical effects which cause dark pulses requires sophisticated metrological methods. In the course of the research presented in this thesis, two innovative methods were developed for the analysis of dark pulses.

The first method is based on the temperature dependence of the dark current of SiPMs. Combining the current measurements at dark conditions and at ambient light conditions, a model for the dark current was developed. This model allows for the extraction of different contributions to the dark current, which (a) depend on the applied voltage, (b) depend on the applied overvoltage and (c) are quasi-independent of the electric field. The characterization of KETEK SiPMs with this method revealed, that the diffusion of charge carriers into the multiplication region is responsible for a major contribution to the dark count rate. The suppression of this effect was achieved by the application of an electric potential to the substrate of the SiPM. By this approach, the dark count rate of KETEK SiPMs was reduced from 300 kHz/mm^2 to 100 kHz/mm^2 at room temperature and an overvoltage of 4 V.

The output signal of SiPMs does not contain any information on the coordinates where the dark pulses have been generated. To overcome this restriction, a novel metrological method was developed within the scope of this work. This method enables a sub-micro-cell, 2D spatially resolved measurement of the dark count rate within the plane of the active area. It is based on the detection and mapping of the light intensity that is emitted by the effect of hot carrier luminescence during the avalanche breakdowns of micro-cells. The relation of the emitted light intensity and the dark count rate is discussed in this thesis. The analysis of the spatially resolved dark count rate revealed the existence of sub-micro-cell regions with a strongly increased charge carrier generation rate (hotspots) due to Shockley-Read-Hall-Generation and trap-assisted tunneling. For KETEK SiPMs, these regions were attributed to crystal defects which are introduced during the dopants implantation in the fabrication process.

The density distribution of crystal defects that is generated during the implantation of the buried n-layer was simulated via the Monte-Carlo method. The obtained results predict, that an increase of the phosphorus implantation energy reduces the density of crystal defects in the active region of micro-cells due to a shift of the phosphorus peak-concentration with respect to the maximum electric field. This prediction was confirmed experimentally, and a further suppression of the total dark count rate down to 40 kHz/mm^2 was achieved.

In a variety of applications, SiPMs are operated in radiation-hostile environments where high-energy particles or γ-rays are causing crystal damages by atomic displacement or electron ionization. The fast increase of the dark count rate with the accumulated irradiation dose sets severe limits to the performance of SiPMs in these kinds of applications. In the course of this project, the impact of the irradiation with ^{60}Co γ-rays and with thermal neutrons on the parameters of KETEK SiPMs was analyzed. In particular, the individual contributions of point defects and cluster defects to the dark count rate were evaluated.

Contents

Chapter 1

Introduction

This thesis has two major points of attention. The first one is the development of innovative characterization methods for the identification and evaluation of the physical effect which underlie the generation of dark pulses in Silicon Photomultipliers (SiPMs). The second one is the reduction of the dark count rate in SiPMs by the suppression of these effects. This chapter introduces the research which is presented in this thesis. In section 1.1, the general idea of single-photon detectors is briefly introduced. Section 1.2 treats the properties of state of the art Silicon Photomultipliers and gives an overview of their field of applications. The motivation of this work and its objectives are presented in section 1.3. In section 1.4 the structure of this thesis is summarized.

1.1   Single photon detection

In general, the setup for the detection of optical signals consists of three major parts: (a) the sensor for the detection of the incoming light, (b) the data acquisition unit for the amplification and readout of the sensor output signal, and (c) the data processing unit for the analysis of the acquired signals. Each of these parts introduces noise to the system which leads to a deterioration of the light detection. If the output signal of the sensor is amplified with a sufficiently large gain, the noise which is introduced by the data processing unit can be neglected. To overcome the limitations imposed by the noise of the electronic amplification circuit on the minimum detectable signal level, optical sensors with an internal amplification are used. Examples of such sensors, which are capable of the detection of single photons, are Photomultiplier Tubes (PMTs), Micro-Channel Plates (MCPs), Hybrid Photon Detectors (HPDs), Single-Photon Avalanche Diodes (SPADs) and Silicon Photomultipliers (SiPMs).

Despite the fundamental differences of the mentioned photon counters, they all share the same general operation principle which is based on the effect of charge-carrier multiplication. The impinging photon is absorbed inside an active volume and generates secondary charge carriers. These charge carriers are accelerated inside the high electric field which is applied across the active volume and gain enough energy to generate further charge carriers along their path, and so on. This multiplication chain leads to an internal gain of the order of 10^5 to 10^7, depending on the photon counter and the operation conditions. For a detailed description of the respective detector principles, the following literature is recommended [1], [2], [3], [4].

1.2 Silicon Photomultiplier

The Silicon Photomultiplier was proposed in the late 1980s by the Russian scientists Z. Sadygov and V. Golovin as an array of Geiger-Mode Avalanche Photodiodes connected in parallel [5], [6]. After a significant progress in the development of SiPMs over the last three decades, they have become promising candidates for a large number of applications which are distributed over a wide range of disciplines like high-energy physics, astro-particle physics, biophotonics, medical applications, dark matter search and many others.

State of the art SiPMs provide a peak photon detection efficiency (PDE) of $40 - 60$ % at a wavelength of $\lambda = 420$ nm. At 550 nm still 60 % of the peak value is achieved [7]. This makes them a suitable candidate for high energy astro-particle physics experiments like MAGIC or EUSO [8]. The application of classical PMTs, which provide a peak PDE of $25 - 35$ % for wavelengths around 400 nm [9], enforces serious limitations on the energy resolution and the accessible energy threshold [10].

A further advantage of SiPMs is, that they can be easily operated at cryogenic temperatures in dark matter search experiments like GERDA [11] or nEXO [12]. In these experiments, SiPMs are used for the detection of the scintillation light of cryogenic noble gases like liquid argon or liquid xenon, which emit in the vacuum ultraviolet (VUV) range ($\lambda = 128$ nm for argon and $\lambda = 175$ nm for xenon). For $\lambda = 128$ nm, state of the art SiPMs with a pitch size of 50 μm show a PDE of approximately 8 % [13]. For $\lambda = 175$ nm, the PDE increases to approximately 15 % [14]. This allows for a direct detection of the scintillation light as done in [12]. The high internal gain of 10^5 to 10^7 allows for the transmission of the output signal over several meters, without amplification. Hence, the operation of a near-detector preamplifier in the cryogenic liquid is not necessary. For this kind of experiments, SiPMs have serious advantages over traditional PMTs. The low operation voltage of $25 - 70$ V of SiPMs compared to $1 - 3$ kV for PMTs, excludes spark breakdowns in noble gas atmosphere, reduces the total power consumption and simplifies the operation of the detector. The small size of SiPMs and the advanced fabrication process of silicon, allows for low concentrations of radioactive impurities which are not achieved in traditional PMTs [15]. The compactness of SiPMs allows for the construction of detectors with a high degree of granularity [16], which is one of its main advantages over traditional PMTs or HPDs for the application in high energy physics (HEP) calorimeters like the CMS HCAL [17]. Since the calorimeters are operated inside a spectrometer magnet, the insensitivity of the SiPMs to magnetic fields is also beneficial for applications in such HEP detectors [18]. One of the weak-points of the SiPMs is the temperature dependence of key parameters like the breakdown voltage and the dark count rate. For the operation of a large number of SiPMs, a control system for parameter monitoring and temperature stabilization is mandatory [19], [20]. In HEP detectors, the SiPMs are operated in radiation-hostile environments. For the CMS HCAL phase I upgrade the accumulated dose of neutrons with energies larger than 100 keV is expected to be approximately $1 - 2 \cdot 10^{12}$ cm^{-2} of 1 MeV equivalent neutrons in the SiPM readout region. For such high irradiation doses, the dark count rate of the SiPMs is expected to increase by several orders of magnitude due to the introduced crystal damages [21]. After a micro-cell has fired, it is not able to detect a photon during its recovery process. The increased dark carrier generation rate after irradiation hence leads to saturation effects and losses in PDE, because the average number of recovering micro-cells at a certain time increases with the dark count rate. In order to increase the radiation hardness of SiPMs, a high dynamic range (large number of micro-cells) and a fast recovery time (small micro-cell capacity) are mandatory. State of the art SiPMs with the highest micro-cell density of 20530 cells/mm^2 and a micro-cell pitch of 7.5 μm were presented by [22]. The recovery time of this SiPM type amounts to approximately 4 ns and the achieved PDE is larger than 20 % at $\lambda = 515$ nm. With a micro-cell pitch of 12.5 μm, a PDE of approximately 30 % at $\lambda = 515$ nm was achieved with a micro-cell density of 7500 cells/mm^2 and a recovery time of approximately 10 ns.

Smaller micro-cells show a faster avalanche buildup time and a faster signal rise time. For this reason, SiPMs with small micro-cells are beneficial for applications which require a high single-photon time resolution [23]. However, the disadvantage of smaller micro-cells is a lower fill factor and hence a lower photon detection efficiency. In medical applications, the "Time-of-Flight Positron Emission Tomography" (TOF-PET) detector-systems impose strong requirements on the SiPM time resolution. TOF-PET systems measure the difference in arrival times of 511 keV annihilation photons at the outer detector ring using a scintillator readout. Here, one of the figures of merit is the coincidence time resolution (CTR) of the detector system, which strongly depends on the properties of the scintillator and the photon detector. CTR values of 205 ps FWHM were reported for a 8x8 LYSO array with a

3

pixel size of 4x4x22 mm^2 matching a SiPM array with an active area of 4x4 mm^2 per pixel [24]. To achieve a high CTR in TOF-PET, the detector has to trigger on the first arriving photons from the scintillation pulse. For this reason, the pulse discriminator threshold must be set at the single photoelectron level. At such low threshold levels, the dark pulses of the SiPM degrade the signal to noise ratio of the detector system and consequently the time resolution [25]. For the selection of events associated with the 511 keV, a good energy resolution is mandatory. As reported in [24], both the coincidence time resolution and the energy resolution improve with a decreasing rate of dark pulses.

It was shown that SiPMs are also suitable for the application in a variety of optical bio-sensors, where the challenge is to detect small photo-signals while maintaining high contrast [26]. In applications like "Fluorescence Correlation Spectroscopy" (FCS) [27] or "Fluorescence Lifetime Imaging" (FLIM) [28], photon detectors with a high time resolution are required, which favors the SiPMs. For applications which use bio-luminescence, additionally the detection of low emission intensities with a high signal to noise ratio is mandatory. Since the minimum detectable light intensity scales with the square root of the dark noise, the relatively high dark count rate of SiPMs with respect to PMTs is a limiting factor for high contrast imaging, especially at room temperature [29]. The high dark count rate also imposes a restriction on the maximum size of the active area of SiPMs. This problem is often solved by coupling the photon detector to optical fibers or micro-lenses, which increases the complexity of the detector system [30]. However, the high dynamic range of SiPMs provides a better linearity over a much broader range of light intensities than PMTs [31]. Further, its compactness, robustness and immunity to damage from light overexposure makes the SiPM a well suited candidate for the realization of compact systems for portable applications.

1.3 Motivation and objectives

The dark count rate represents the intrinsic noise limit of the Silicon Photomultiplier and restricts the performance of this photon detector in many applications.

The creation of charge carriers at dark conditions may occur by a variety of physical mechanisms. The ratio of these mechanisms may vary, depending on the SiPM design, fabrication technology and operation conditions. For this reason, there is no general solution providing a low dark count rate for every SiPM technology. Every developer has to individually identify the mechanisms which are dominating the dark count rate of the respective SiPM. The fact that the output signal of the SiPM does not provide any spatial information on its origin makes the separation of the contributing mechanisms even more challenging.

Therefore, an important objective of the research presented in this thesis is to develop suitable methods for the investigation of physical mechanisms which determine the dark count rate of Silicon Photomultipliers.

Over the last years, several producers achieved a reduction of the dark count rate to a level of 30 to 100 kHz/mm^2. On the contrary, the KETEK SiPMs still show a dark count rate which is approximately one order of magnitude higher with respect to the state of the art devices (see figures 1.1(a) and 1.1(b)). For this reason, the second goal of this research is to make the KETEK SiPM a competitive photon detector by finding a way to suppress the mechanisms which are dominating its dark count rate.

In applications, in which the SiPMs are operated in radiation-hostile environments, the dark count rate is the parameter which degrades the fastest due to the generated defects in the silicon crystal. One of the requirements to SiPMs in these applications is to provide a dark count rate, which increases as slowly as possible with the accumulated irradiation dose, rather than being initially low. The third goal of this thesis is to contribute to the development of the radiation hardness of Silicon Photomultipliers by evaluating the dark count rate of point and cluster defects generated during the irradiation with ^{60}Co γ-rays and with thermal neutrons.

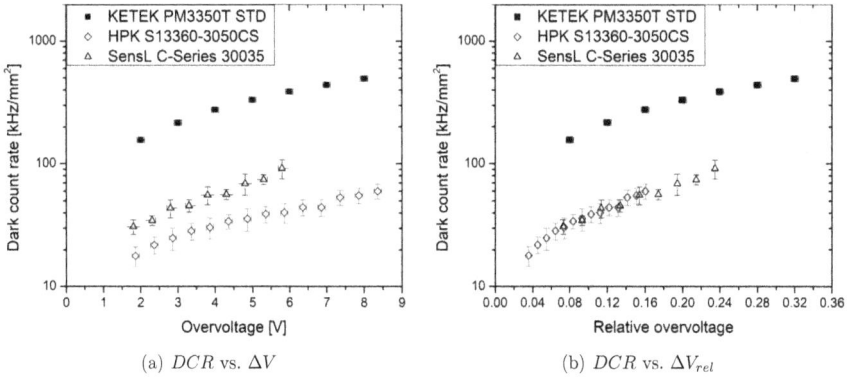

(a) DCR vs. ΔV (b) DCR vs. ΔV_{rel}

Figure 1.1: Comparison of the dark count rate of SiPMs from different producers at $T = 21\ ^{\circ}C$.

1.4 Organization of this thesis

This thesis describes the research that was done in order to understand the origin of dark pulses in KETEK SiPMs and find suited solutions for their suppression. The thesis is divided into 9 chapters, of which this introduction is the first.

The **second chapter** treats the fundamentals of Silicon Photomultipliers. This chapter begins with a review of the operation principle of the SiPM, based on its equivalent circuit. In the second part of this chapter, the two main physical processes of impact ionization and avalanche multiplication are presented and their influence on the photon detection efficiency of the SiPM is discussed. The last part reviews the physical mechanisms which are potentially contributing to the noise of the SiPM.

In the **third chapter**, the experimental methods used for the determination of selected SiPM parameters are presented. In the first part of this chapter, the experimental setup is described on which all other setups used in this work are based. Furthermore, the developed algorithm for the signal processing and pulse detection is introduced. The second part of this chapter is focused on the determination of the SiPM parameters which are important for the line of argument in this thesis. Special attention is paid to the temperature dependence of these parameters. The last part of chapter three summarizes the SiPMs which were used in the presented research.

Chapter four discusses the identification of the diffusion current as the strongest contributor to the generation of dark pulses in KETEK SiPMs. The temperature dependence of the dark current of the SiPMs is used as an indicator of physical processes. The activation energies of these processes are extracted from the Arrhenius plots of the dark current. In the first part of this chapter, the conventional method for the extraction of the activation energies is applied and its drawbacks are discussed. In the second part of this chapter, a novel method is presented which was developed in the course of this project in order to overcome these drawbacks and improve the confidence level of the obtained results.

In **chapter five**, the elaborated solution for the suppression of the dominating diffusion current in KETEK SiPMs is presented. It is based on the application of an electric potential to the substrate of the photon detector which counteracts the diffusion of charge carriers.

Chapter six presents a method for the spatially resolved characterization of the dark count rate of Silicon Photomultipliers. The method is based on the idea of mapping the light intensity, which is emitted by the effect of hot carrier luminescence during the avalanche breakdowns of micro-cells. In the first part of this chapter, the existence of regions with an enhanced dark count rate is discussed and the contribution of these hotspots to the total dark count rate of KETEK SiPMs is experimentally determined. In the second part of this chapter, the areas in which the substrate potential has an impact on the dark count rate are identified and the achieved suppression of dark pulses is evaluated in these areas. The last part of this chapter presents an approach for the generation of activation energy maps for a spatially resolved characterization of physical mechanisms which contribute to the dark count rate of SiPMs. Using this approach, the contribution of the remaining diffusion current to the dark count rate of KETEK SiPMs is evaluated. Furthermore, the hotspots are identified as crystal defects which are introduced by the phosphorus implantation in the fabrication process.

Chapter seven treats the optimization of the phosphorus implantation parameters, aiming for a reduced contribution of hotspots to the dark count rate of KETEK SiPMs. The chapter begins with the simulation of the implantation defects for two distinct implantation energies. The simulation results are confirmed experimentally in the second part of this chapter and SiPMs with a further suppressed dark count rate are presented. In the last part of this chapter, the variation of the hotspot density and the variation of the dark count rate is investigated.

In **chapter eight**, the degradation of KETEK SiPMs due to radiation damages is characterized. The impact of the irradiation with ^{60}Co γ-rays and with thermal neutrons is evaluated. Additionally, the dark count rate which is generated by point defects is compared with the one generated by cluster defects. This chapter introduces an interesting and important field of research and motivates further projects headed in this direction.

Finally, **chapter nine** summarizes the results of the research which is presented in this thesis and gives an outlook on possible future research projects.

Chapter 2

Fundamentals

In this chapter, the fundamental physical processes are reviewed on which the operation characteristics of the SiPM are based. The chapter is divided in three main parts.

In the first part (section 2.1), the operation principle of the SiPM is presented and the equivalent circuit is discussed.

The second part of this chapter (sections 2.2 - 2.4) reviews the fundamental processes of impact ionization and avalanche multiplication. Their influence on the most crucial SiPM parameter, the photon detection efficiency, is discussed.

In the third part of this chapter (section 2.5 - 2.8), the physical origins of dark pulses, optical crosstalk and afterpulsing are presented. These processes are responsible for the generation of false pulses during the operation of the SiPM and hence deteriorate the detection of incoming photons.

2.1 Operation principle of the SiPM

A Silicon Photomultiplier is a semiconductor device that provides the ability to detect single photons with a high gain of the order of 10^5 to 10^7. The device consists of an array of avalanche photodiodes (APD) connected in parallel and operated above their breakdown voltage (V_{BD}), in Geiger-Mode (GM-APD). In this work, the individual GM-APDs are also referred to as micro-cells of the SiPM. A sketch of a GM-APD is shown in figure 2.1. It consists of p-doped and n-doped silicon layers. The layers can be generated by implantation or diffusion of dopants into the silicon crystal. Also epitaxial growth techniques are used. At the junction of these layers (pn-junction) a high internal electric field is build up, when the device is operated at reverse bias conditions. If an electron-hole pair (e-h pair) is generated inside the high-field region of the GM-APD, the electron is dragged towards the n-contact (cathode), while the hole is dragged towards the p-contact (anode).

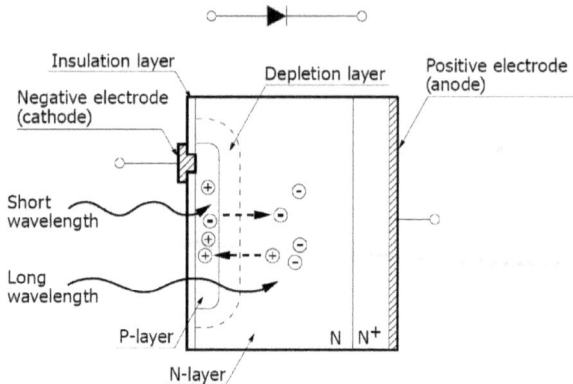

Figure 2.1: Sketch of a Geiger-Mode avalanche photodiode (GM-APD). Taken from [1].

If the electric field exceeds a certain threshold, charge carriers are able to initiate an avalanche breakdown by the mechanisms of impact ionization. As a consequence, a breakdown current is flowing, which is large enough to be detected (see sections 2.2 and 2.3). In a GM-APD, the applied electric field strength is sufficiently high for electrons and holes to impact ionize. For this reason, the charge carrier avalanche is self sustaining and must be quenched by reducing the voltage across the diode below the breakdown voltage, after the detection of a breakdown. For the SiPMs investigated in this work, the quenching is achieved by a voltage drop across a serial polysilicon resistor (in the following referred to as quenching resistor R_q). However, also active quenching circuits may be used.

The generation of an e-h pair may occur by the absorption of a photon with sufficient energy, which leads to the single photon detection capabilities of the SiPM. However, also e-h pairs which are not generated by photons lead to avalanche breakdowns and consequently to false signals (see section 2.5). In figure 2.2, the equivalent circuit diagram of a SiPM is shown, as proposed by [32]. The model consists of a parallel connection of N avalanche photo diodes, with one of them firing at a time. This corresponds to the case of one dark pulse without correlated pulses. The depletion region of the pn-junction is modeled by the capacitance C_d. Additionally, every micro-cell consists of a polysilicon quenching resistor R_q in parallel to the parasitic capacitance C_q. The series resistance of the micro-plasma in the avalanche is accounted for by R_d [33]. The charge released by one micro-cell during an avalanche breakdown is modeled by the current source I_{AV}. The parasitic capacitance introduced by the metal interconnects of the micro-cells is accounted for by the grid capacitance C_g.

Considering a typical metal-to-substrate capacitance per unit area of 0.03 fF/μm^2 and a fill factor of roughly 65 %, C_g of the KETEK PM3350T STD (chip size of 3x3 mm^2, micro-cell size of 50x50 μm^2) is estimated by 95 pF, whereas the total micro-cell capacity $(C_q + C_d)$ of the PM3350T STD amounts to 220 fF.

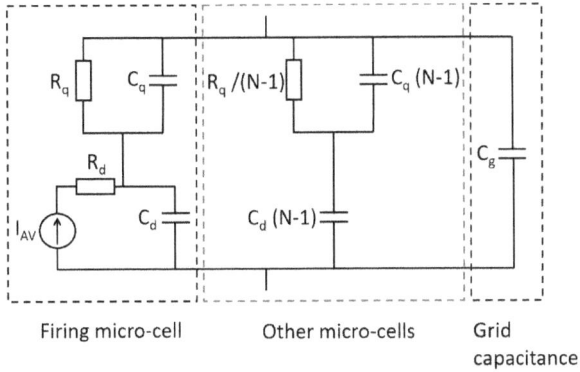

Figure 2.2: Equivalent circuit diagram of a SiPM.

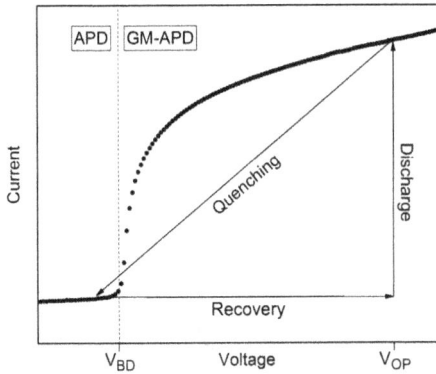

Figure 2.3: Operation principle of a SiPM.

In figure 2.3, the operation principle of a SiPM is shown. It can be subdivided into three steps:
(i) Discharge: The SiPM is operated in the Geiger-Mode at the operation voltage $V_{OP} > V_{BD}$. In this mode, the device is in a quasi-stable condition, at which no current is flowing. If a free charge carrier is generated in the high-field region, a discharge of the micro-cell via an avalanche breakdown is initiated with a finite probability. The rise time of the corresponding current pulse can be approximated by equation 2.1 [33]. The total charge Q released during a breakdown of one micro-cell is given by equation 2.2. Here, the quantity ΔV is called overvoltage or excess bias voltage.

$$\tau_{rise} \approx R_d \cdot (C_q + C_d) \tag{2.1}$$

$$Q = (C_q + C_d) \cdot (V_{OP} - V_{BD}) = (C_q + C_d) \cdot \Delta V \tag{2.2}$$

(ii) Quenching: During the Geiger-discharge, the voltage drop over the quenching resistor reduces the voltage at the junction below V_{BD}. The electric field drops below the impact ionization threshold and the charge carrier avalanche is quenched.
(iii) Recovery: The recovery of the micro-cell consists of two components. The first component is determined by the rapid charge supply of the parasitic capacitance C_q [34]. For the KETEK PM3350T STD, C_q contributes to about 10 % of the total charge output and consequently amounts to approximately 22 fF. The recovery time of the fast component can be estimated by equation 2.3. Here, C_{eq} represents the series connection of $(N-1) \cdot C_d$ and $(N-1) \cdot C_q$ [34]. R_{load} is the input resistance of the front-end electronics.

$$\tau_{fast} = R_{load} \cdot (C_{eq} + C_g) \tag{2.3}$$

The second recovery component is determined by the recharge of the micro-cell through R_q with the characteristic time constant τ_{slow}, which is given in equation 2.4. During the recovery, the electric field inside the multiplication region of the micro-cell exceeds the ionization threshold level and the micro-cell is ready to be discharged again.

$$\tau_{slow} = R_q \cdot (C_q + C_d) \tag{2.4}$$

In figure 2.4, the measured and the simulated pulse shape of the PM3350T STD are compared. The simulation is based on the equivalent circuit from figure 2.2 and was performed with the LTspice simulator [35]. The resistance R_d was neglected. How measured pulse shape was recorded will be discussed in section 3.3.

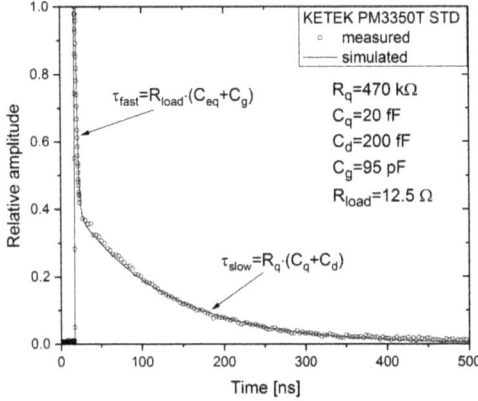

Figure 2.4: Comparison of the simulated and the measured pulse shape of the KETEK PM3350T STD.

If two or more photons simultaneously trigger avalanche breakdowns in different micro-cells, the output signal of the SiPM is a superposition of the individual micro-cell signals and the released charge is proportional to the number of firing micro-cells. However, if two or more photons are simultaneously absorbed in one micro-cell, the released charge is equivalent to the firing of one micro-cell. For this reason, the number of firing micro-cells and hence the total charge output Q_{tot} is in general not proportional the number of incident photons. The response of the output signal can be approximated by using equation 2.5, with N_{fired} being the number of triggered micro-cells, N_{cells} being the total number of micro-cells of the SiPM, N_{ph} being the number of incident photons and PDE being the photon detection efficiency of the SiPM [36], [1].

$$Q_{tot} \propto N_{fired} = N_{cells} \left(1 - \exp\left[-\frac{N_{ph} \cdot PDE}{N_{cells}} \right] \right) \tag{2.5}$$

At the condition that $(N_{ph} \cdot PDE)$ is small compared to N_{tot}, the response of the output signal is approximately linear with the number of incident photons. The maximum number of photons for which this condition is fulfilled is called dynamic range. For a fixed SiPM size, a trade-off has to be made between the photon detection efficiency and the dynamic range (see section 2.4). In figure 2.5(a) an overlay of amplified signal pulses of the PM3350T STD are shown. Here, the SiPM was illuminated with a pulsed laser and the SiPM output

13

was synchronized with the laser output. The corresponding pulse height spectrum is shown in figure 2.5(b). The unit "p.e." stands for "photo-electron" and represents the number of fired micro-cells.

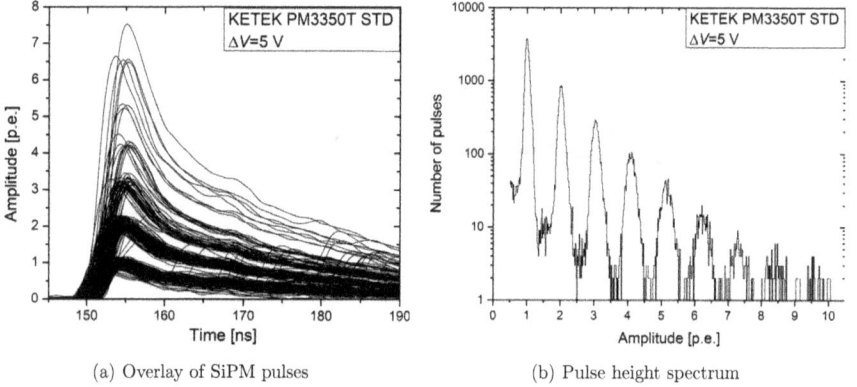

(a) Overlay of SiPM pulses

(b) Pulse height spectrum

Figure 2.5: Amplified signal pulses and pulse-height spectrum of the PM3350T STD at $\Delta V = 5$ V.

2.2 Impact ionization

If a SiPM is operated at reverse biased conditions, free electrons and holes in the depletion region travel against or in direction of the electric field towards the n-layer or p-layer respectively. If the applied electric field is high enough, the charge carriers may acquire enough energy from the field to ionize a silicon crystal atom when colliding. Such ionizing collisions lead to the generation of secondary electron-hole pairs (e-h pairs). This process is called impact ionization. The minimum energy required for impact ionization is called the ionization threshold energy E_i. This material dependent quantity is different for electrons ($E_{i,e}$) and holes ($E_{i,h}$). In table 2.1, the ionization threshold energies along the principal axes of silicon are shown [37].

Table 2.1: Ionization threshold energies

Crystal orientation	< 100 >	< 111 >	< 110 >
$E_{i,e}$	1.1 eV	3.1 eV	2.1 eV
$E_{i,h}$	1.8 eV	2.9 eV	1.8 eV

To describe the probability that a charge carrier, which gained sufficient energy, will actually impact ionize, the concept of ionization coefficients for electrons (α_e) and holes (α_h) is used. The ionization coefficient is defined as the average number of secondary e-h pairs which are generated by a charge carrier, traveling a unit distance. This quantity is determined by the energy of the charge carriers. For low electric field gradients, a direct correlation of the ionization coefficients and the electric field is assumed.

A variety of models are existing which try to describe the ionization coefficients in terms of the applied electric field. In the following, two of them are briefly presented:

Wolff [38] determined the ionization rate by considering the optical phonon scattering as the dominating charge carrier decelerating process that competes with impact ionization. All other scattering mechanisms were not considered. The assumption was made, that the energy gained by a charge carrier between two collisions is larger than the energy loss in an optical phonon scattering event. Additionally it was assumed that the mean free path for impact ionization is much smaller than the mean free path for optical phonon scattering. In this model, the charge carrier constantly gains energy, until it exceed the ionization threshold. The resulting expression for the ionization coefficient is given in equation 2.6. Here, F is the electric field strength. $A_{e,h}$ and $b_{e,h}$ are material constants valid for electrons and holes respectively.

$$\alpha_{e,h}(F) = A_{e,h} \cdot \exp\left(-\frac{b_{e,h}}{F^2}\right) \tag{2.6}$$

In a competing model, it is assumed that the energy gain between two collisions is less than the energy loss during a phonon scattering event. In contrast to the previous model, charge carriers cannot constantly gain energy to reach the ionization threshold. Only those few charge carriers, which avoid collisions are able to reach the threshold energy in order to impact ionize a silicon atom. At this boundary conditions, Shockley [39] derived the following expression for the field dependency of the ionization coefficient:

$$\alpha_{e,h}(F) = A_{e,h} \cdot \exp\left(-\frac{b_{e,h}}{F}\right) \tag{2.7}$$

For most experiments, a better description of the experimental data is achieved by using equation 2.7 [40]. In table 2.2, the ionization coefficients measured by van Overstraeten and de Man are shown. In figure 2.6, α_e and α_h are calculated as a function of F by using equation 2.7. For low electric field strengths, α_e is much larger than α_h. However, the ratio of the ionization coefficients decreases with increasing F. This has an impact on the avalanche triggering probability and hence on the design of SiPMs (see section 2.4).

Table 2.2: Ionization coefficients. Taken from [41]

	Electrons	Holes	Holes
$F\ [10^5\ \mathrm{Vcm^{-1}}]$	$1.75 \leq F \leq 6.0$	$1.75 \leq F \leq 4.0$	$4.0 \leq F \leq 6.0$
$A_{e,h}\ [10^6\ \mathrm{cm^{-1}}]$	0.703	1.582	0.671
$b_{e,h}\ [10^6\ \mathrm{Vcm^{-1}}]$	1.231	2.036	1.693

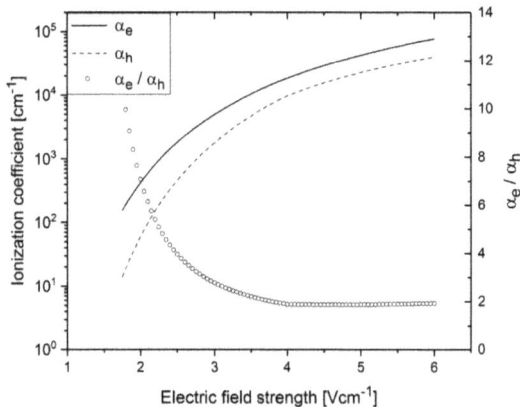

Figure 2.6: Ionization coefficients for electrons and holes. The calculation was performed using equation 2.7 and the values from table 2.2.

2.3 Avalanche multiplication and breakdown

In the previous section, the generation of secondary electron-hole pairs by impact ionization was discussed. In this section, the build up of a charge carrier avalanche in a reverse biased

16

pn-junction is described. In figure 2.7, a schematic of such a junction is shown. If an e-h pair is generated at the position x within the junction, the electron will be accelerated towards the n-contact, while the hole will be accelerated toward the p-contact. While traveling a distance dx, the electron will generate an average number of $\alpha_e dx$ secondary e-h pairs. Similarly, the hole will suffer $\alpha_h dx$ ionizing collisions, while traversing the distance dx. The secondary e-h pairs will themselves generate additional e-h pairs, which will result in a large number of ionization chains. The average total number of generated e-h pairs, starting from an initial e-h pair generated at x, is defined as the gain G.

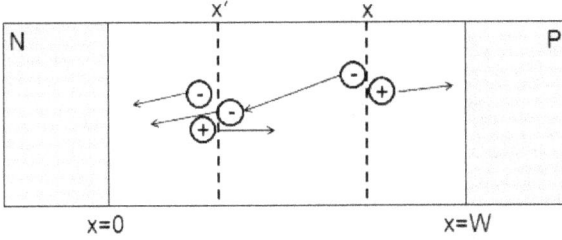

Figure 2.7: Reverse biased pn-junction. Electron which is generated at x, generates a secondary e-h pair at x'.

In the following, the derivation of G is given following the calculations by McIntyre [42]. The initial equation that describes the considerations above, is given in 2.8. This expression points out that every e-h pair generated at x' in an ionization chain resulting from an initial e-h pair generated at x, will itself be multiplied on the average by $G(x')$.

$$G(x) = 1 + \int_0^x \alpha_e G(x') \mathrm{d}x' + \int_x^W \alpha_h G(x') \mathrm{d}x' \qquad (2.8)$$

The derivative of equation 2.8 with respect to x is given by 2.9. This differential equation is solved by 2.10.

$$\frac{\mathrm{d}G(x)}{\mathrm{d}x} = (\alpha_e - \alpha_h)\, G(x) \qquad (2.9)$$

$$G(x) = G(0) \cdot \exp\left[\int_0^x (\alpha_e - \alpha_h)\, \mathrm{d}x'\right] = G(W) \cdot \exp\left[-\int_x^W (\alpha_e - \alpha_h)\, \mathrm{d}x'\right] \qquad (2.10)$$

In order to determine the average gain for the case that the ionization chain is initiated by an electron generated at the p-layer edge ($x = W$), equation 2.10 is substituted in equation 2.8 for $x = W$. The resulting expression for $G(W)$ is given in equation 2.11.

$$G(W) = \frac{1}{1 - \int_0^W \alpha_e \exp\left[-\int_x^W (\alpha_e - \alpha_h)\,\mathrm{d}x'\right]\,\mathrm{d}x} \qquad (2.11)$$

Inserting equation 2.11 into equation 2.10 results in expression 2.12 for the gain $G(x)$.

$$G(x) = \frac{\exp\left[-\int_x^W (\alpha_e - \alpha_h)\,\mathrm{d}x'\right]}{1 - \int_0^W \alpha_e \exp\left[-\int_{x'}^W (\alpha_e - \alpha_h)\,\mathrm{d}x''\right]\,\mathrm{d}x'} \qquad (2.12)$$

The condition for the occurrence of an avalanche breakdown is fulfilled when the avalanche multiplication gain becomes infinite. This is the case if the ionization integral, which is given in equation 2.13, becomes one and consequently the denominator in equation 2.12 becomes zero. The ionization integral is often used in TCAD simulators (Technology Computer Aided Design) to determine the breakdown voltage. The operation voltage is increased in steps. For every step the ionization coefficients are determined from the electric field distribution. Then the ionization integral is calculated. Depending on whether the ionization integral is larger or smaller than one, the operation voltage is increased or decreased.

$$\int_0^W \alpha_e \exp\left[-\int_{x'}^W (\alpha_e - \alpha_h)\,\mathrm{d}x''\right]\,\mathrm{d}x' \stackrel{!}{=} 1 \qquad (2.13)$$

2.4 Photon detection efficiency

One central parameter of SiPMs is the photon detection efficiency PDE. It is defined as the number of detected photons divided by the number of incident photons. In equation 2.14, the photon detection efficiency is described by a product of three parameters:

$$PDE = \varepsilon \cdot QE \cdot P_{trigg} \qquad (2.14)$$

(i) The fill factor ε is the fraction of the SiPM area, which is able to detect photons. The area of the SiPM which is not sensitive to light is mainly due to the metal lines for signal readout, the quenching resistors, guard rings for electric field attenuation towards the micro-cell edges and trenches for the suppression of optical crosstalk. One way of maximizing PDE without changing the general micro-cell structure, is to increase the micro-cell size and consequently to increase ε. However, the dynamic range will be decreased by this approach. Another possibility is to use transparent materials for the quenching resistor to maximize the active area [43].

(ii) The quantum efficiency QE describes the probability that a photon is absorbed within the active region of a micro-cell and generates an e-h pair. One factor which determines the quantum efficiency is the reflection and refraction at the silicon surface. In order to maximize the photon transmittance, anti-reflective coatings (ARC), which consist of several layers of silicon oxide and silicon nitride, are deposited onto the Si surface [44]. The coatings have to be optimized for the spectral range of the detected light. After the photon has passed the ARC, it has to be absorbed within the active region of the micro-cells in order to be detected. The absorption of a photon can be described by the Beer-Lambert law, which is shown in equation 2.15. Here, I_0 is the photon flux at the surface and α is the absorption coefficient, which is a function of the wavelength λ.

$$I(x) = I_0 \cdot \exp[-\alpha x] \tag{2.15}$$

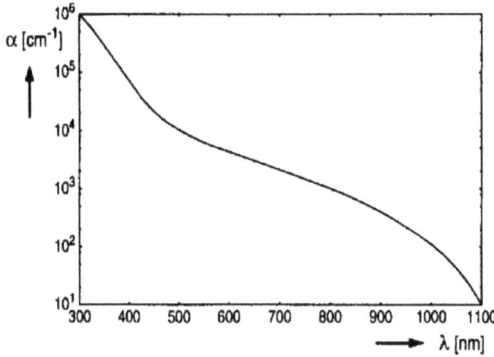

Figure 2.8: Photon absorption coefficient α vs. wavelength, in silicon. Taken from [40].

In figure 2.8, α is plotted versus λ for the light absorption in silicon. For photon energies larger than the bandgap energy, the photon can be absorbed by a transition of an electron from the valence band to the conduction band and hence a generation of an e-h pair. Because silicon is an indirect bandgap material, an additional phonon is needed to fulfill momentum conservation. With increasing photon energy, less phonon momentum is required and α is consequently increasing. Direct transitions without the participation of a phonon become available for photon energies higher than 3.4 eV. For this reason there is a further increase of the absorption coefficient for wavelengths smaller than 365 nm [40]. Photons with a short λ are absorbed close to the Si surface. To maximize PDE for such photons, a shallow dopants implantation of the top layer is mandatory. In this way the depletion region is extended as

19

close to the surface as possible. Lower energetic photons are absorbed at larger depths. To reach an enhanced detection of such photons, the active region has to reach deeper inside the silicon. This can be achieved for example by a higher implantation energy, or epitaxial grown layers. However, with the increasing thickness of the pn-junction, the electric field increases slower with the applied voltage. For this reason the breakdown voltage is shifted to higher values and the maximum PDE is reached at higher absolute overvoltages.

(iii) The avalanche triggering probability P_{trigg} describes the probability that a generated e-h pair will successfully initiate an avalanche breakdown by impact ionization. The avalanche triggering probability is a function of the impact ionization coefficient $\alpha_{e,h}$ and is hence different for electrons and holes (see section 2.2). The different probabilities can be described by the differential equations 2.16 and 2.17 [45]. Here, $P_{e,h}(x)$ is the probability that an electron/hole starting from the position x initiates an avalanche breakdown while traveling the distance dx. The total probability that an avalanche is triggered by an e-h pair is given in equation 2.18.

$$\frac{dP_e}{dx} = (1 - P_e) \cdot \alpha_e \cdot (P_e + P_h - P_e P_h) \tag{2.16}$$

$$\frac{dP_h}{dx} = -(1 - P_h) \cdot \alpha_h \cdot (P_n + P_h - P_n P_h) \tag{2.17}$$

$$P_{trigg} = P_e + P_h - P_e P_h \tag{2.18}$$

For a pn-junction which extends from $x = 0$ within the n-layer to $x = W$ within the p-layer, the equations are solved by a numerical integration over the depletion region [46]. The boundary conditions that have to be fulfilled are $P_e(0) = 0$ and $P_h(W) = 0$, since electrons which are generated at the non-depleted n-layer edge will leave the depletion regions without causing an ionization. The same condition holds for holes at the non-depleted p-layer edge. In figure 2.9(a), the solutions of equations 2.16 and 2.17 are shown as a function of the position at which the charge carriers are generated. The results correspond to an avalanche photo-diode with $V_{BD} \approx 27$ V.

In figure 2.9(b), the calculated and experimentally determined avalanche triggering probabilities are compared for electrons and holes which are generated as minority carriers at the edge of the depletion region. The experimental data was obtained by illuminating a diode with short (390 nm) and long (1050 nm) wavelengths [46].

The obtained results reveal that the probability of triggering an avalanche is much higher for electrons than for holes. Furthermore, P_e reaches saturation at much lower overvoltages than P_h. Consequently the structure of the SiPM has to be designed such that electrons are used to trigger the avalanche breakdown. This means that a large fraction of the incident light should be absorbed close to the p-layer. To maximize PDE for blue light a p-on-n structure is favorable, whereas an n-on-p structure is preferentially sensitive to red light.

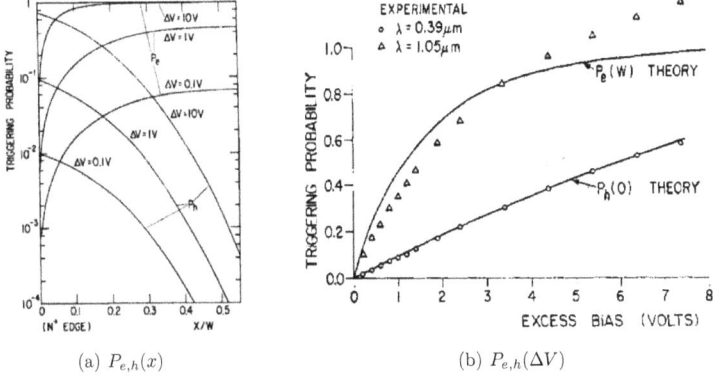

(a) $P_{e,h}(x)$ (b) $P_{e,h}(\Delta V)$

Figure 2.9: (a) Triggering probabilities $P_{e,h}$ at several overvoltages as a function of the position of the e-h pair generation. (b) $P_{e,h}$ for electrons/holes starting as minority carriers at the edge of the depletion region. Taken from [46].

2.5 Mechanisms responsible for dark pulses

The breakdown of a GM-APD is initiated by the generation of an electron-hole pair within the depletion region. For this reason, e-h pairs that are not produced by photon absorption, but by other mechanisms, will also lead to an avalanche breakdown. Such events are referred to as dark counts or dark pulses. In this section, the main mechanisms responsible for the occurrence of dark pulses in Silicon Photomultipliers are discussed. In literature, five main mechanisms are reported. One mechanism is the diffusion of minority charge carriers from quasi neutral regions into the multiplication region [47], [48]. A further mechanism is the generation of electron-hole pairs by means of the Shockley-Read-Hall-Generation due to defects in the depletion region [1], which can be enhanced by a high electric field leading to the mechanism of trap assisted tunneling or the Poole-Frenkel effect. For higher electric field strengths, direct band-to-band tunneling contributes to the generation of electron-hole pairs, as reported in [49]. Which of the named mechanisms is dominating, depends on the device geometry, electric field strength and operation temperature.

21

2.5.1 Diffusion of charger carriers

The diffusion process originates from a gradient of the charge carrier concentration. The charge carriers are migrating from regions of higher concentrations to regions of lower concentrations according to Fick's first law, as shown in equation 2.19. Here, n and p are the electron and hole concentrations, $J_{n,p}$ is the electron or hole current density and $D_{n,p}$ is the diffusion coefficient for electrons or holes. As a consequence of the diffusion current, an electric field F builds up, which results in a counteracting drift current density. The obtained total current density equation is accordingly changed to 2.20, with $\mu_{n,p}$ being the electron or hole mobility.

$$J_{n,p} = qD_{n,p}\nabla(n,p) \tag{2.19}$$

$$J_{n,p} = J_{drift} + J_{diffusion} = qn\mu_{n,p}F \pm qD_{n,p}\nabla(n,p) \tag{2.20}$$

In order to obtain the time dependence of the current densities, the continuity equation (see equation 2.21) is considered. It describes the change of the charge carrier concentration by the net generation/recombination rate U and the net currents flowing in and out of the observed region (J_{drift}, $J_{diffusion}$).

$$\frac{\partial(n,p)}{\partial t} = -U \pm \frac{1}{q}\nabla J_{n,p} \tag{2.21}$$

Considering steady-state conditions for the n-region of the junction, the one dimensional continuity equations for electrons and holes can be written as shown in equations 2.23 and 2.22. Here, n_n and p_n are the electron and hole concentrations at the n-side of the pn-junction.

$$-U - \mu_p F\frac{dp_n}{dx} - \mu_p p_n\frac{dF}{dx} + D_p\frac{d^2p_n}{dx^2} = 0 \tag{2.22}$$

$$-U + \mu_n F\frac{dn_n}{dx} + \mu_n n_n\frac{dF}{dx} + D_p\frac{d^2n_n}{dx^2} = 0 \tag{2.23}$$

Due to charge neutrality ($n_n - n_{n0} - p_n + p_{n0} = 0$), it follows that $dn_n/dx = dp_n/dx$. Here, n_{n0} and p_{n0} describe the equilibrium electron and hole densities at the n-side of the pn-junction. Multiplying equation 2.22 with $\mu_n n_n$ and equation 2.23 with $\mu_p p_n$ and applying the additional condition of low-level injection at the n-side of the junction ($p_n << n_n \approx n_0$), equations 2.22 and 2.23 can be reduced to equation 2.24. Here, τ_p is the carrier recombination lifetime for holes, which depends on the hole capture cross-section σ_p, the density of bulk traps N_t and the thermal velocity v_{th} with the charge carrier effective mass m^*.

$$-\frac{p_n - p_{n0}}{\tau_p} - \mu_p F \frac{dp_n}{dx} + D_p \frac{d^2 p_n}{dx^2} = 0 \tag{2.24}$$

$$\tau_{n,p} = \frac{1}{\sigma_{n,p} v_{th} N_t} \tag{2.25}$$

$$v_{th} = \sqrt{\frac{3kT}{m^*}} \tag{2.26}$$

Considering neutral regions, where the electric field is zero, equation 2.24 can be further reduced to:

$$\frac{d^2 p_n}{dx^2} - \frac{p_n - p_{n0}}{D_p \tau_p} = 0 \tag{2.27}$$

This differential equation is solved by 2.30 as described in [50]. The applied boundary condition are shown in equations 2.28 and 2.29. Here, W_n is the depletion width on the n-side of the pn-junction.

$$p_n(W_n) = p_{n0} \exp\left(\frac{qV}{kT}\right) \tag{2.28}$$

$$p_n(x = \infty) = p_{n0} \tag{2.29}$$

$$p_n(x) - p_{n0} = p_{n0} \left[\exp\left(\frac{qV}{kT}\right) - 1\right] \exp\left[-\frac{x - W_n}{\sqrt{D_p \tau_p}}\right] \tag{2.30}$$

By the combination of equation 2.30 and Fick's first law 2.19, the diffusion current at $x = W_n$ reveals the well known Shockley equation for holes:

$$J_p = -qD_p \frac{dp_n}{dx}\Big|_{W_{Dn}} = J_{p0}\left[\exp\left(\frac{qV}{kT}\right) - 1\right] \tag{2.31}$$

$$J_{p0} = qp_{n0}\sqrt{\frac{D_p}{\tau_p}} \approx q\sqrt{\frac{D_p}{\tau_p}} \frac{n_i^2}{N_D} \tag{2.32}$$

Since the Shockley equation represents the ideal diode law, it has to be modified slightly to be applicable to describe the dark pulses of the Silicon Photomultiplier due to diffusion. If the minority charge carrier reaches the multiplication region, it will initiate an avalanche breakdown only with a certain probability $P_{trigg}^{diffusion}$. For this reason, 2.32 changes to:

$$J_{p0} \approx q\sqrt{\frac{D_p}{\tau_p}} \frac{n_i^2}{N_D} \cdot P_{trigg}^{diffusion} \tag{2.33}$$

In order to determine the temperature dependence of the diffusion current, the temperature dependence of all terms in equation 2.32 is considered.

At high doping concentrations and low temperatures, the mobility μ is limited by the ionized impurity scattering mechanism and shows the temperature dependence $\mu(T) \propto T^{\frac{3}{2}}$ [50]. For lower doping concentrations and $T > 100$ K, the mobility is dominated by acoustic phonon scattering, with the resulting temperature dependence $\mu(T) \propto T^{-\frac{3}{2}}$ [50], [51]. Using the Einstein relation 2.34, the temperature dependence of the diffusion coefficient is determined to be $D(T) \propto T^{\frac{5}{2}}$ or $D(T) \propto T^{-\frac{1}{2}}$.

$$D = \frac{kT}{q}\mu \tag{2.34}$$

The temperature dependence of the intrinsic charge carrier density n_i is given by equation 2.35 [52], where the bandgap energy is also a function of T. $E_g(T)$ follows the semi empirical Varshni equation 2.36, with $E_g(0) = 1.1692$ eV, $\alpha = (4.9 \pm 0.2) \cdot 10^{-4}$ eV/K and $\beta = (655 \pm 40)$ K [53].

$$n_i \propto T^{1.6} \exp\left(-\frac{E_g(T)}{2kT}\right) \tag{2.35}$$

$$E_g(T) = E_g(0) - \frac{\alpha T^2}{T + \beta} \tag{2.36}$$

Neglecting the temperature dependence of the hole capturing cross section σ_p in 2.25, the hole recombination lifetime shows a temperature dependence of $\tau_p(T) \propto T^{-\frac{1}{2}}$. Combining all terms, the temperature dependence of the diffusion current density can be written as shown in equation 2.37.

$$J_{diffusion} \propto T^{3.2} \exp\left(-\frac{E_g(T)}{kT}\right) \tag{2.37}$$

Neglecting the power law term in T with respect to the exponential temperature dependence, the activation energy for the diffusion mechanism, determined from the slope of the Arrhenius plot $(\ln(J_{diffusion})$ vs $1/kT)$ will be mainly defined by the energy gap E_g.

2.5.2 Shockley-Read-Hall-Generation and trap-assisted tunneling

The Shockley-Read-Hall (SRH) generation/recombination process has its origin in the retrieval of the equilibrium condition $p \cdot n = n_i^2$. For the boundary conditions of $p \cdot n > n_i^2$ the recombination of charge carriers is dominating. The generation of charge carriers becomes dominant for $p \cdot n < n_i^2$. For a reverse biased pn-junction, the free charge carrier concentration in the depletion region is lowered $(p \cdot n << n_i^2)$. Consequently, the dominant SRH process is the one of charge carrier generation, where impurities with energy levels inside the bandgap act as generation centers for electron-hole pairs. A generated charge carrier inside the high-field region is then able to initiate an avalanche breakdown by the process of impact ionization.

The net generation/recombination rate U is described by the Shockley-Read-Hall equation 2.38. As already mentioned, the term $pn - n_i^2$ determines whether a net recombination or generation is dominating. For impurities with energy levels within the bandgap, U reaches its maximum at $E_t = E_i$, which means that generation/recombination centers with energy levels close to mid-bandgap show an enhanced contribution the the net transition rate. Under the boundary condition of a reverse biased pn-junction, the free electron and hole concentrations are lowered ($p << n_i$ and $n << n_i$) and equation 2.38 can be reduced to 2.40. Here, τ_n and τ_p are the electron and hole recombination lifetimes. The rate at which electron-hole pairs are generated inside the depletion region is described by the charge carrier generation lifetime τ_g [54].

$$U = \frac{\sigma_p \sigma_n v_{th} N_t (pn - n_i^2)}{\sigma_n \left[n + n_i \exp\left(\frac{E_t - E_i}{kT} \right) \right] + \sigma_p \left[p + n_i \exp\left(-\frac{E_t - E_i}{kT} \right) \right]} \qquad (2.38)$$

$E_i :=$ Fermi level of intrinsic semiconductor

$E_t :=$ energy level of crystal lattice impurity

$\sigma_{n,p} :=$ electron and hole capture cross-section

$N_t :=$ concentration of impurity states

$v_{th} :=$ thermal velocity

$k :=$ Boltzmann constant

$T :=$ temperature $\qquad (2.39)$

$$U \approx \frac{-n_i}{\tau_n \exp\left(\frac{E_t - E_i}{kt} \right) + \tau_p \exp\left(-\frac{E_t - E_i}{kt} \right)} \equiv -\frac{n_i}{\tau_g} \qquad (2.40)$$

In order to determine the generation current density I_{gen}, equation 2.40 is multiplied by the avalanche triggering probability for generation P_{trigg}^{gen} and integrated over the depletion width W. The result in shown in equation 2.41. P_{trigg}^{gen} may differ from $P_{trigg}^{diffusion}$, since both quantities are functions of the electric field and consequently depend the position at which the charge carrier is created or at which the charge carrier enters the region of impact ionization. Approximating $P_{trigg}^{gen}(x)$ with a constant effective avalanche triggering probability $\bar{P}_{trigg}^{gen}$, the expression for the pure SRH generation current can be simplified further.

$$J_{gen} = \int_0^W q \cdot |U| \cdot P_{trigg}^{gen} \, dx \approx \frac{q n_i W}{\tau_g} \cdot \bar{P}_{trigg}^{gen} \tag{2.41}$$

In order to determine the temperature dependence of the pure SRH generation current, two conditions are imposed. The first one is that only impurities/traps with $E_t \approx E_i$ are considered, since they are the most dominant contributors to the generation rate. The second condition is that the temperature dependence of $\bar{P}_{trigg}^{gen}$ can be neglected. This approximation was experimentally confirmed by measuring the photon detection efficiency as a function of temperature (see section 3.5). Under these conditions, the temperature dependence of I_{gen} is determined by the intrinsic charge carrier concentration n_i and the charge carrier generation lifetime τ_g. It can be expressed as shown in equation 2.42.

$$J_{gen} \propto T^{2.1} \exp\left(-\frac{E_g}{2kT}\right) \tag{2.42}$$

Neglecting the power law term in T with respect to the exponential temperature dependence, the activation energy for the pure SRH generation current, determined from the slope of the Arrhenius plot ($\ln(J_{gen})$ vs $1/kT$) will be mainly defined by the mid-bandgap energy gap $E_g/2$. Equation 2.40 is valid at low electric field strengths. After the charge carrier is captured by a trap, the emission to the conduction band occurs over the full trap depth $E_t - E_c$, with E_c being the energy level of the conduction band. The electric field strength in the depletion region of a GM-APD is of the order of 10^5 V/cm. In the presence of such a high electric field, the thermal emission is enhanced by the trap-assisted tunneling. In this process, the charge carrier is emitted only over a part of the trap depth, followed by a phonon-assisted tunneling process to the conduction band. This mechanism is schematically shown in figure 2.12. In [49], this effect was modeled by dividing the electron and hole recombination lifetimes in equation 2.40 by the field-effect function $(1+\Gamma)$, which results in the following expression for the generation rate:

$$U \approx \frac{-n_i}{\frac{\tau_n}{1+\Gamma} \exp\left(\frac{E_t - E_i}{kt}\right) + \frac{\tau_p}{1+\Gamma} \exp\left(-\frac{E_t - E_i}{kt}\right)} \equiv -\frac{(1+\Gamma)\, n_i}{\tau_g} \tag{2.43}$$

For an electric field strength $F < 9 \cdot 10^5$ Vcm, Γ can be expressed as shown in equations 2.44 and 2.45, with m_t being the effective mass for tunneling, which equals to 0.25 times the free electron mass m_0 [49].

$$\Gamma = 2\sqrt{3\pi} \frac{|F|}{F_\Gamma} \exp\left[\left(\frac{F}{F_\Gamma}\right)^2\right] \tag{2.44}$$

$$F_\Gamma = \frac{\sqrt{24m_t\,(kT)^3}}{q\hbar} \tag{2.45}$$

Analogous to equation 2.42, the temperature dependence of the enhanced generation current $J_{gen+tat}$ can be expressed as shown in equation 2.46. As a consequence of the assumed constant effective avalanche triggering probability $\bar{P}_{trigg}^{gen}$, the electric field F is also replaced by the effective field strength F_{eff}.

$$J_{gen+tat} = J_0 \cdot T^{2.1} \exp\left(-\frac{E_g}{2kT}\right) \cdot \left(1 + \frac{2\sqrt{3\pi}|F|}{F_\Gamma} \cdot \exp\left[\frac{q^2\hbar^2 F_{eff}^2}{24m_t\,(kT)^3}\right]\right) \tag{2.46}$$

For boundary conditions of a negligible electric field, equation 2.46 reduces to the Shockley-Read-Hall expression 2.42. For non-negligible electric field strengths, the activation energy for the generation of electron-hole pairs is reduced according to the last term of equation 2.46. In figure 2.10, a simulated Arrhenius plot of $J_{gen+tat}/J_0$ is shown between $T = +40°C$ and $T = -40°C$. For rather low electric fields $(F_{eff} \approx (1-5) \cdot 10^5$ V/cm), the activation energy can be estimated from the slope of a linear fit to the data points. For higher electric fields $(F_{eff} \gtrsim 5 \cdot 10^5$ V/cm), the $(kT)^{-3}$ dependence of $J_{gen+tat}$ becomes dominant and $J_{gen+tat}/J_0$ cannot be approximated with a single exponential function of $1/kT$.

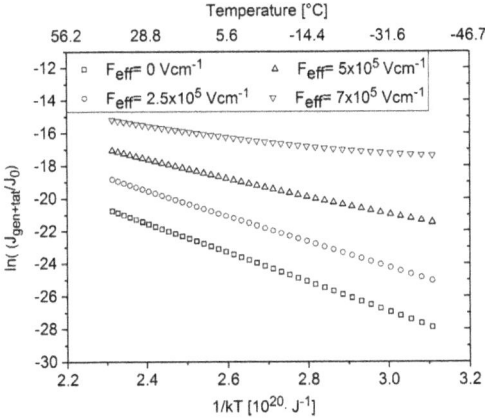

Figure 2.10: Simulated Arrhenius plots of the field-enhanced SRH generation current at different effective electric field strengths.

In figure 2.11, the lowering of the predicted activation energy is shown for effective electric field strengths F_{eff} between 0 V/cm and $5.5 \cdot 10^5$ V/cm.

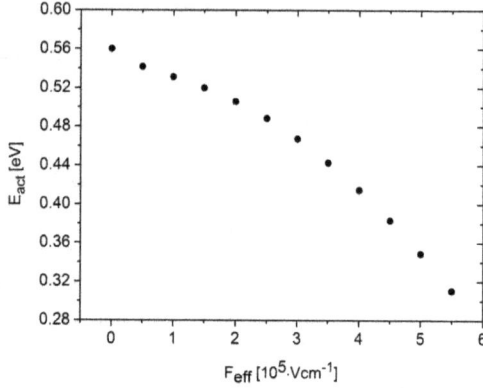

Figure 2.11: Simulated activation energy of the field-enhanced SRH generation current as a function of the effective field strength.

2.5.3 Poole-Frenkel effect

The Poole-Frenkel effect is based on the Coulomb interaction between the free charge carrier and the impurity or trap, due to which the trap depth is lowered and the emission rate of the charge carrier is enhanced. In contrast to the phonon-assisted tunneling, which is independent of the charge state of the trap, the Poole-Frenkel effect requires a charged impurity. The electric field and temperature dependence of the dark current induced by the increased charge carrier emission due to the Poole-Frenkel effect is described in equation 2.47 [55].

$$J_{PF} \propto \exp\left(\frac{\delta_{PF}}{kT}\right) \; ; \; \text{with } \delta_{PF} = \sqrt{\frac{Zq^3F}{\varepsilon}} \tag{2.47}$$

Here, Z is the charge of the trap, q is the elementary charge, F is the electric field strength and ε is the dielectric constant. Equation 2.47 predicts that an Arrhenius plot of the Poole-Frenkel induced dark current ($\ln(J_{PF})$ vs $1/kT$) would show a linear increase with a slope proportional to the square root of the electric field ($\sqrt{F}$). The observed activation energy of the generation current would be further reduced in the presence of a non-negligible Poole-Frenkel effect. In figure 2.12, the fundamental principles of the Poole-Frenkel effect, the trap-assisted tunneling and the direct band-to-band tunneling are shown schematically.

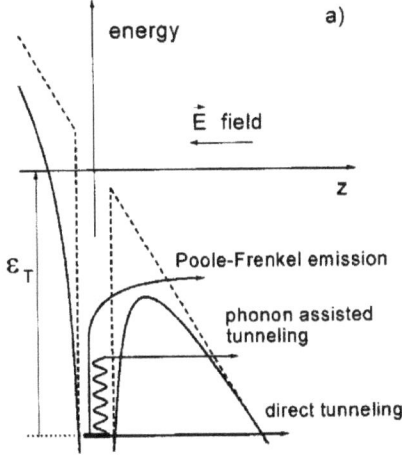

Figure 2.12: Sketch of the fundamental principles of the Poole-Frenkel effect, the trap-assisted tunneling and the band-to-band tunnelin. Taken from [55]

2.5.4 Band-to-band tunneling

At high electric field strengths, an electron-hole pair may also be generated by a direct transition of an electron from the valence band to the conduction band. With an increasing electric field strength, the width of the barrier between the valence and the conduction band decreases, which is schematically shown in figure 2.13. Consequently, the probability density of an electron in the conduction band increases. Hurkx et al. [56] modeled the band-to-band tunneling current as a Dirac δ-function generation term at the location of the maximum electric field. The obtained expression for the band-to-band current density J_{bbt} is given in equation 2.48.

$$J_{bbt} = c \cdot V_{reverse} \cdot \left(\frac{F_{max}}{F_0} \right)^{\frac{3}{2}} \exp \left(-\frac{F_0}{F_{max}} \right) \cdot P_{trigg}^{max} \qquad (2.48)$$

Here, c is a constant, $V_{reverse}$ is the reverse bias voltage at the junction and F_{max} is the maximum electric field strength. F_0 is a material constant and depends on the form of the potential barrier. For an assumed parabolic barrier, F_0 can be written as 2.49 [40].

$$F_0 = \frac{\pi \sqrt{m_{eff} E_g^{\frac{3}{2}}}}{2\sqrt{2} q \frac{h}{2\pi}} \qquad (2.49)$$

29

Assuming that tunneling predominantly occurs at the position of the maximum electric field at the junction, the avalanche triggering probability P_{trigg}^{max} is used in equation 2.48. If the temperature dependence of the bandgap energy E_g is neglected, the band-to-band tunneling rate is independent of T.

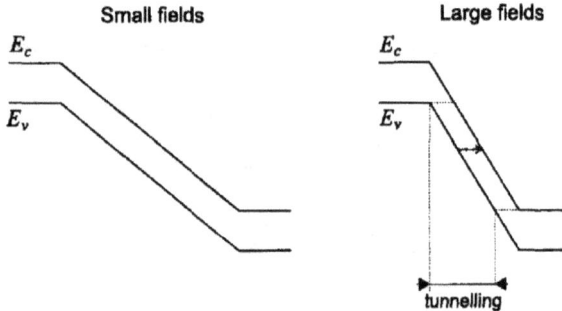

Figure 2.13: Sketch of the decreasing of the bandgap barrier with an increasing electric field strength. Taken from [40]

2.6 Hot carrier luminescence

When an avalanche breakdown occurs, light is emitted from the active volume of a micro-cell. This effect is referred to as electro-luminescence or hot carrier luminescence. The physical origin of the photon emission process is still not consistently understood. In literature, various mechanisms are proposed for the explanation of the energy spectrum of the emitted light. The fact that the spectra reported by different authors do not show a similar behavior, reveals the controversy concerning this topic. For Silicon-Photomultipliers, the number and energy of secondary photons are crucial parameters. After the breakdown of one micro-cell, the emitted photons can penetrate into neighboring micro-cells and initiate additional avalanche breakdowns. This effect can result in direct or delayed crosstalk pulses, which depends on the penetration path of the photon and the region of its absorption inside the pn-junction. Mirzoyan et al. [57] reported an emission of $2.6 \cdot 10^{-5}$ photons per avalanche electron in a wide spectral range from 500 nm to 1600 nm. In figure 2.14, the reported differential light emission spectrum for a $1mm^2$ Hamamatsu MPPC $S10362 - 11 - 10U$ is shown.

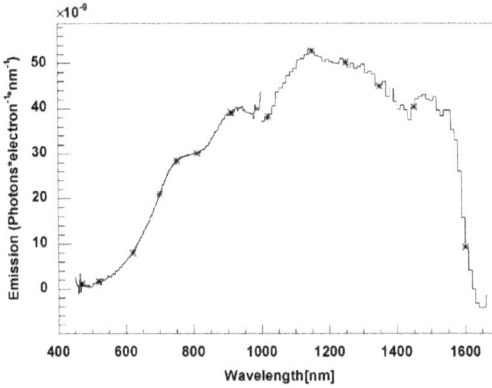

Figure 2.14: Measured differential light emission spectrum for a 1 mm^2 Hamamatsu MPPC $S10362 - 11 - 10U$. Taken from [57]

2.6.1 Mechanisms of light generation in an avalanche breakdown

Light emission from reverse biased pn-junctions was already reported by Newman et al. in 1955 [58]. In order to describe and model the light emission spectra from avalanche breakdowns, three main mechanisms were proposed by Akil et al. [59]: (1) ionization and indirect interband transitions, (2) direct interband transitions and (3) Bremsstrahlung. In figure 2.15 the proposed model is compared with an experimentally determined spectrum from a reverse biased pn-junction [59]. In the energy range below 2 eV the indirect interband transition model proposed by Gautam et al. [60] is applied in order to fit the experimental data. In this model, the emission intensity is expressed as a product of the recombination probability and a high-field distribution function which includes acoustical, optical and ionizing scattering.

During the avalanche discharge, the carriers gain energy from the applied electric field and are ionized by impact ionization in the presence of acoustical and optical phonon scattering. During this process, the ionized charge carriers return to lower energy states (recombination) by radiative indirect transitions involving the interaction with a phonon for the purpose of momentum conservation. The energy of a charge carrier which is not transmitted to a phonon is lost by the emission of light.

For photon energies above 2.3 eV, the direct interband model proposed by Wolff [61] provides the best fit to the experimental data. In direct interband transitions, the relaxation of the energetic charge carrier occurs without a change in momentum and consequently does

31

not require an interaction with a phonon. In figure 2.16, the radiative direct and indirect conduction band to valence band transitions are shown schematically.

For photon energies between 2 eV and 2.3 eV the bremsstrahlung is reported to be the predominant mechanism, which is the breaking radiation of hot electrons in the Coulomb field at a charged impurity [62]. Bremsstrahlung is a particular kind of intraband transitions of hot electrons (holes) in the conduction (valence) band. The electron scatters on a charged impurity transferring its momentum to the impurity by a simultaneous energy transfer to an emitted photon. In this kind of transitions, only one type of charge carriers is involved in contrast to the above described interband transitions. The hot electron (hole) relaxes from an excited energy state to a lower energy state within the same energy band.

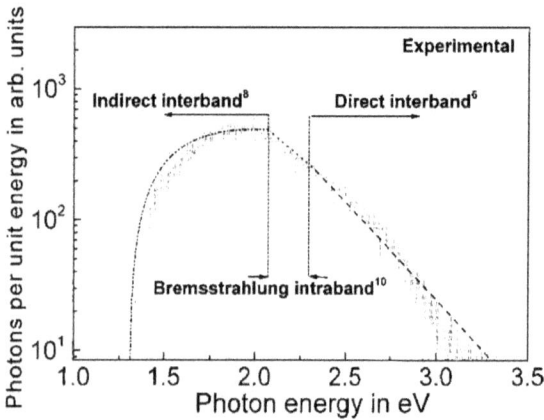

Figure 2.15: The multi-mechanism model proposed by Akil et al. is compared with experimental data. Taken from [59].

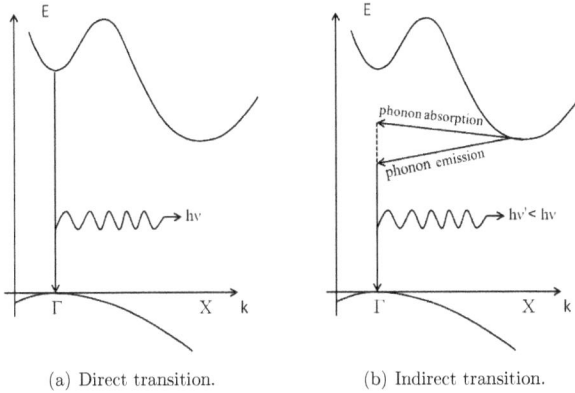

(a) Direct transition.

(b) Indirect transition.

Figure 2.16: Sketch of the direct and indirect transitions from the conduction band to the valence band for silicon.

Lacaita et al. [63] reported an emission of $2.9 \cdot 10^{-5}$ photons with energy higher than 1.14 eV per carrier crossing the reverse biased pn-junction. The energy spectrum published in [63] (see figure 2.17) significantly differs from the spectrum reported in [59] (see figure 2.15). The authors in [63] observed a linear increase of the experimentally determined photon rate with the reverse dark current of the avalanche photodiode (see figure 2.18). This result excludes the contribution of electron-hole recombinations, since it would lead to a square power dependence of the photon rate on the avalanche current. Bude et al. [64] reported that the contribution of bremsstrahlung is negligible for impurity densities $< 5 \cdot 10^{20}$ cm^{-3}, in silicon. For SiPMs which were investigated in this work, such high dopant concentrations are not reached. According to this considerations, electron (hole) energy relaxations between states in the conduction (valence) band are expected to be the dominating mechanism for the effect of hot carrier luminescence.

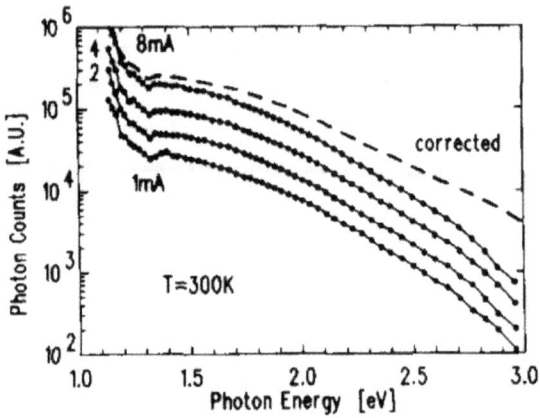

Figure 2.17: The photon emission spectra of a reverse biased APD at room temperature at different reverse currents. Taken from [63].

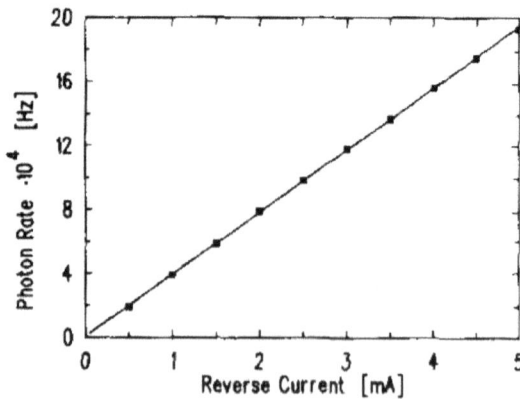

Figure 2.18: The hot carrier induced photon emission rate of an APD as a function of the reverse current. Taken from [63].

2.7 Prompt and delayed optical crosstalk

As discussed in the previous section, optical photons are generated in the charge carrier avalanche by a variety of processes. These photons are able to propagate to neighboring micro-cells and initiate further avalanche breakdowns. The propagation of the photons occurs by several paths, as depicted in figure 2.19. One scenario is that an emitted photon is absorbed directly or after several reflections in the high field region of a neighboring micro-cell and initiates an avalanche breakdown. In this case, the time difference between the first and the consecutive pulse is < 1 ns [65]. This is not sufficient for a distinction of the two pulses. For this reason, only one pulse with an amplitude of 2 p.e. is registered. This effect is called "prompt optical crosstalk" (CT). In order to suppress this effect, trenches which are filled with opaque materials (e.g. Aluminum, Tungsten, etc.) are implemented around the borders of micro-cells [10], [66], [67], [68]. The significant suppression of optical crosstalk by the implementation of trenches was reported in [69].

Another scenario is that the emitted photon is absorbed beneath the high field region. The generated minority charge carriers then diffuse to the active region. This process is significantly slower compared to the prompt optical crosstalk and is hence called "delayed optical crosstalk" (DCT). The consecutive breakdowns can be distinguished and are registered as multiple pulses with an 1 p.e. amplitude [68]. For the suppression of the DCT, a second pn-junction at the backside of the micro-cell can be used. For KETEK SiPMs, this junction is formed by the p-type substrate and the n-layer, as shown in figure 2.19. The additional potential barrier is supposed to prevent hole diffusion from the bulk to the active region [67].

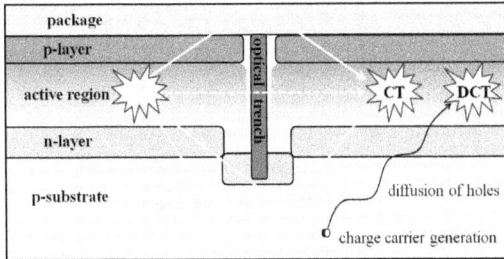

Figure 2.19: Sketch of possible photon paths leading to prompt optical crosstalk (CT) and delayed optical crosstalk (DCT).

The optical crosstalk probability scales with the number of generated photons during an avalanche breakdown, the geometric cross-section for the interaction between two micro-cells and the photon detection efficiency of the micro-cells [70].

35

For a variety of applications, optical crosstalk is an undesirable effect. It deteriorates the linearity of the output signal and complicates the correlation of the measured number of photons to the incident one [71]. In PET (Positron Emission Tomography), an enhanced crosstalk probability clearly deteriorates the energy resolution by artificially increasing the detection efficiency [72] or leading to saturation effects [73]. A detailed analysis of this deterioration is given in [74].

2.8 Afterpulsing

During an avalanche breakdown, a large number of charge carriers flows through the depletion region of the pn-junction. Energy states present within the bandgap may capture electrons or holes from the conduction or valence band and re-emit them after a certain delay-time Δt into the same band. Such energy states are called trapping centers or traps and are caused by crystal defects and impurities. If the trap is located in the active region of a micro-cell, the re-emitted charge carrier has a finite probability to trigger a subsequent avalanche in the same micro-cell. Such events are called afterpulses. The delay-times depend on the respective trap type and may vary by many orders of magnitudes [40].

The probability to observe an afterpulse in the time interval dt after the primary avalanche breakdown occurs at $t = 0$, is given by equation 2.50 [75]. Here, τ_i is the characteristic time constant of the re-emission process of the i-th trap. The factor $p_i(t)$ describes the probability that a charge carrier is captured during an avalanche breakdown and triggers an afterpulse when re-emitted. $p_i(t)$ is a function of t, since the avalanche triggering probability increases with the electric field strength which itself depends on the recovery state of micro-cell.

$$P_{AP}(t)dt = \sum_i \frac{p_i(t)}{\tau_i} \exp\left(-\frac{t}{\tau_i}\right) dt \qquad (2.50)$$

In figure 2.20, the amplitude of pulses subsequent to primary dark pulses are plotted versus the delay-time Δt. The amplitude of an afterpulse A_{AP} which occurs at Δt after the primary dark pulse, depends on the recovery state of the micro-cell at that time, and is consequently a function of the recovery time τ_{rec}. For small Δt, afterpulses and dark pulses can be distinguished by their amplitude. Fitting equation 2.51 to the data points attributed to afterpulses, the recovery time of the micro-cells can be determined [76] (see dashed line in figure 2.20).

$$A_{AP}(\Delta t) = 1 \text{ p.e.} \cdot \left(1 - \exp\left[-\frac{\Delta t}{\tau_{rec}}\right]\right) \qquad (2.51)$$

For short delay-times, the detection of consecutive pulses becomes inefficient, which explains the missing data points at short Δt. The critical delay-time Δt_{crit} for which two pulses can still be distinguished strongly depends on the applied pulse shaping. Due to their larger amplitudes at short delay-times, Δt_{crit} of delayed crosstalk pulses is shorter than the one of afterpulses. In the presented example, afterpulses are detected at $\Delta t > 20$ ns, whereas delayed-crosstalk pulses are detected at $\Delta t > 10$ ns. For SiPMs with a shorter recovery time, Δt_{crit} of afterpulses is decreased.

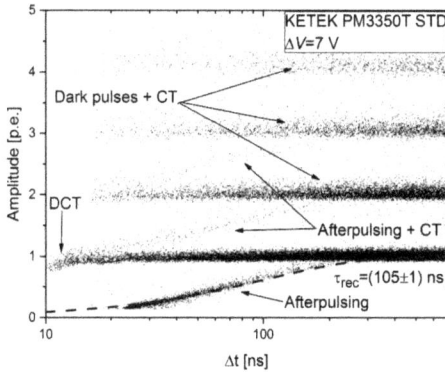

Figure 2.20: Amplitudes of pulses subsequent to primary dark pulses vs. their delay-time.

Chapter 3

Determination of Standard SiPM Parameters

In this chapter, the methods which were applied in this work for the characterization of standard SiPM parameters are presented. Except for the current-voltage characteristics, the parameters were determined via a signal trace analysis. The SiPM signal traces were first recorded at different voltage and temperature conditions. Afterwards, the data set was analyzed offline with a custom algorithm written in LabVIEW and C++. Only those SiPM parameters are discussed, which are essential for the line of argument in this work.

3.1   Experimental setup

In this section, the fundamental experimental setup is discussed, which was used as a basis for every measurement presented in this work. Detailed descriptions of the individual setups are presented in the respective chapters.

In order to prevent the illumination of the SiPM with ambient light, all measurements were performed in a dark box. For the bias supply and current measurements, the *Keithley 6487 piccoammeter/voltage source* or the *Keithley 4200 Parameter Analyzer* were used. The output signal of the SiPM was connected to the *Photonique, AMP-0611* transimpedance amplifier with a gain of 10. Depending on the investigated sample, the connection was realized via pins or probe needles. The *PSI, DRS4 Evaluation Board* was used for the measurement of the dark count rate, of the photon detection efficiency, of the optical crosstalk probability, of the probability of correlated pulses, as well as for the generation of pulse-height spectra. The measurement of the absolute gain of the SiPM was performed with the *LeCroy, Waverunner 64 MXi-A Oscilloscope*.

3.2 Signal processing and pulse detection

The algorithm for the detection of pulses which is presented in this work, is equivalent to a leading edge discriminator (LED). A pulse is detected if the signal amplitude exceeds the threshold level T_{det}. The time of the transition of T_{det} is considered as the occurrence time of the pulse. The limited sampling rate of the acquisition introduces an uncertainty to the timing analysis. In order to minimize the time jitter of the threshold transition, the sampling points are linearly interpolated and the intersection with T_{det} is calculated.

If a subsequent pulse occurs within the recovery of the primary pulse, a pile-up is observed. In this case, the application of an LED is not possible. Based on the approach reported in [77], the filter function F^k_{MWD} is applied to the recorded signal traces in order to eliminate the long recovery tail. The abbreviation MWD stands for "Moving Window Difference". The application of the filter function to the n-th signal point with the amplitude A_n, results in the difference $(A_n - A_{n-k})$. The formal definition of the filter principle is given in equation 3.1. With respect to the original signal trace, the length of the filtered signal trace is reduced by k samples. In this work, k was set to 6. Because the signal traces were acquired with a sampling rate of 1GS/s, the first 6 ns of each signal trace were lost.

$$F^k_{MWD}(A_n) = A_n - A_{n-k} \qquad (3.1)$$

In order to reduce the arbitrary noise, which is not attributed to the SiPM signal, the filter function F^m_{MWA} was applied. Here, the abbreviation MWA stands for "Moving Window Average". The application of F^m_{MWA} to the n-th signal point with the amplitude A_n, results in an average amplitude of the n-th and $(m-1)$ preceding data points. The formal definition of this filter principle is given in equation 3.2. Due to the averaging of the signal, the amplitude of the SiPM pulses is reduced. However, this does not effect the characterization of the SiPM parameters which are discussed in this chapter. The applied methods are based on the ratio of the pulse amplitudes, which is invariant under F^m_{MWA}. Another undesirable effect of the averaging filter is that the minimal time difference of two pulses Δt_{crit} at which a distinction is possible, increases with the increasing of the parameter m. In this work, m was set to 4. The consecutive application of F^6_{MWD} and F^4_{MWA} limits Δt_{crit} to 10 ns, see figure 2.20.

$$F^m_{MWA}(A_n) = \frac{1}{m} \sum_{i=0}^{m-1} A_{n-i} \qquad (3.2)$$

In figure 3.1, a part of a recorded signal trace of the PM3350T STD is shown at $T = 40\ °C$ and $\Delta V = 5$ V. The high operation temperature leads to an enhanced DCR and consequently to a frequent pile-up of pulses. The application of the MWD filter effectively reduces the pulse lengths due to the removed exponential tail and creates a stable baseline which allows for a pulse detection with an LED. Evidently, this approach allows for a stable reconstruction of the amplitudes of piled up pulses and hence the generation of pulse-height spectra for the SiPM characterization.

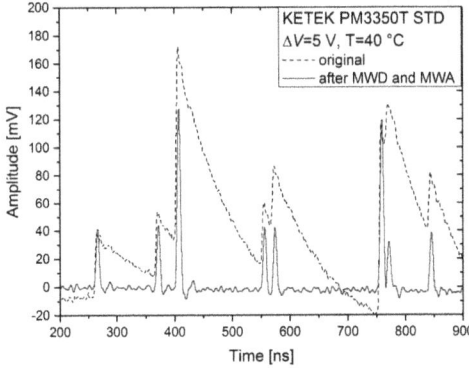

Figure 3.1: Example of a signal trace of a PM3350T STD before and after the application of the MWD and MWA filters.

3.3 Gain and recovery time

The gain G of the SiPM is determined by the charge Q released during an avalanche break-down of one micro-cell divided by the elementary charge e. Q can be expressed as the product of the overvoltage ΔV and the micro-cell capacity C_{cell} [78].

$$G = \frac{Q}{e} = \frac{\Delta V \cdot C_{cell}}{e} \tag{3.3}$$

In this work, the gain was measured by flashing the SiPM with a pulsed laser (406 nm, 1 kHz repetition rate, < 70 ps pulse length). The intensity of the laser output was chosen such that the SiPM was driven into saturation with every light pulse (number of emitted photons >> number of micro-cells). Figure 3.2 shows the schematic of the used experimental setup and figure 3.3 shows a recorded pulse of the PM3350T STD at $\Delta V = 5$ V.

3. Determination of Standard SiPM Parameters

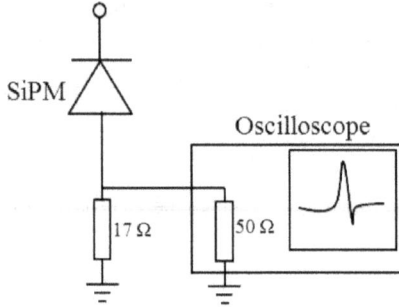

Figure 3.2: Sketch of the experimental setup for the gain and the recovery time measurement.

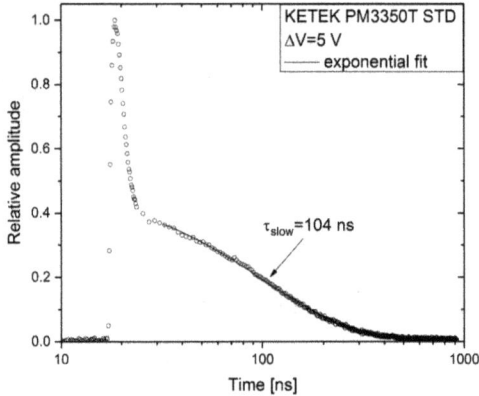

Figure 3.3: Pulse shape of the saturated PM3350T STD (all micro-cells fire simultaneously).

For each flash, the SiPM signal was integrated for a duration of 500 ns ($\approx 5 \cdot \tau_{rec}$) and divided by the load resistance R_{load}. This resulted in the total released charge during the breakdown of N_{cells} micro-cells. The gain was then determined by dividing the integral charge by the total number of SiPM micro-cells N_{cells} and the elementary charge.

$$G = \frac{\int_0^{500 \text{ ns}} V \, dt}{R_{load} \cdot N_{cells} \cdot e}, \tag{3.4}$$

From an exponential fit to the data points in figure 3.3, the time-constant of the slow recovery component was determined to amount to $\tau_{slow} = 104$ ns. This result is in agreement with the time-constant determined in figure 2.20. Using equation 2.3, the time-constant of the fast recovery component was estimated by $\tau_{fast} \approx 2$ ns. In the remainder of this work, τ_{fast} is neglected and τ_{slow} is referred to as the recovery time τ_{rec}.

The presented method is applicable due to the small variation of the pulse shapes of the individual micro-cells. The advantage of this method is, that the uncertainties due to the electronic noise are minimized. Additionally, no amplifier is required, which may introduce a shaping of the SiPM signal. A disadvantage of this method is, that the contribution from afterpulsing, which increases the charge output and the recovery time, is not taken into account. However, for the SiPMs investigated in this work, the afterpulsing probability can be neglected, as will be discussed in section 3.7.

3.4 Breakdown voltage

If the applied bias voltage exceeds a certain threshold, enough charge carriers are multiplied and form a plasma channel. Under this condition, the Geiger discharge is turned on [79]. This threshold voltage is called the turn-on breakdown voltage V_{BD}^{on}. Two approaches are reported in [79] to measure V_{BD}^{on}. The first approach is to determine the voltage at which the photon detection efficiency drops to zero. Since the measurement of PDE is only possible above the breakdown voltage, the measured data points have to be fitted with an appropriate model. The second approach is to determine the voltage at which the multiplied dark current of the SiPM ($I_{dark}(V \geq V_{BD})$) drops to zero. In this approach, it is assumed that the voltage dependence of the multiplied dark current can be described by a power law as shown in equation 3.5 [80]. This assumption is reasonable, since the parameters by which I_{dark} is determined (see equation 4.2) show a linear or quadratic increase with ($V - V_{BD}^{on}$).

$$I_{dark}(V \geq V_{BD}) \propto (V - V_{BD}^{on})^n \tag{3.5}$$

The inverse logarithmic derivative (ILD) of I_{dark} with respect to V is given in equation 3.6. The turn-on breakdown voltage V_{BD}^{on} is defined as the intercept of the linear fit to $ILD(I_{dark})$ in the vicinity of the breakdown voltage with the voltage axis. In figure 3.4, I_{dark} and $ILD(I_{dark})$ are shown for the PM3350T STD at $T = 21\ ^\circ$C.

$$ILD(I_{dark}) = \left[\frac{d}{dV}\ln(I_{dark})\right]^{-1} = \frac{(V - V_{BD}^{on})}{n} \tag{3.6}$$

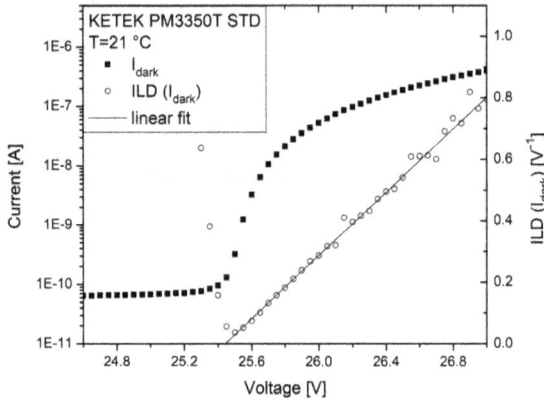

Figure 3.4: Dark current of the PM3350T STD at $T = 21\ °C$ and the corresponding ILD.

In contrast to the turn-on breakdown voltage, also the quantity called the turn-off breakdown voltage V_{BD}^{off} exists. It is defined as the threshold voltage at which an already existing plasma channel is maintained [79]. A common approach to determine V_{BD}^{off} is to measure the gain G as a function of the applied voltage. The gain increases linearly with V and drops to zero at $V = V_{BD}^{off}$. The intercept of the linear fit of $G(V)$ with the voltage axis gives V_{BD}^{off}. In figure 3.5(a), $G(V)$ is shown for the PM3350T STD.

In this work, V_{BD}^{off} was determined by a modification of this method. Instead of the gain G, a quantity called the relative gain G_{rel} was used. G_{rel} is defined as the difference between the 1 p.e. and the 2 p.e. amplitude. For all SiPMs which are discussed in this work, G_{rel} is proportional to G (see figure 3.5(b)). For this reason, the turn-off breakdown voltage does not depend on whether the gain G or the relative gain G_{rel} is used for the measurement.

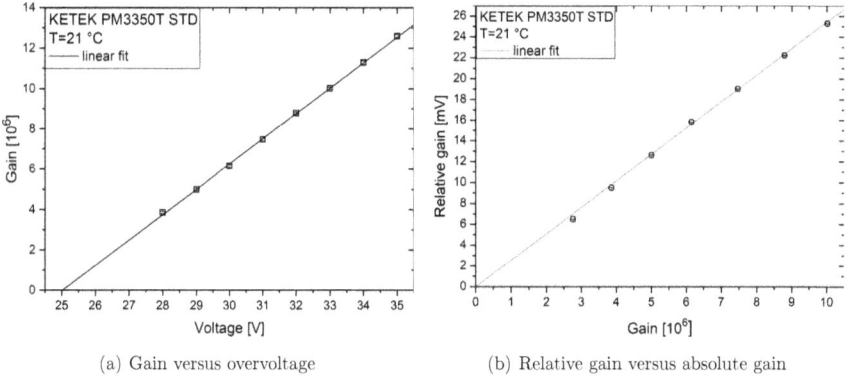

(a) Gain versus overvoltage

(b) Relative gain versus absolute gain

Figure 3.5: (a) Linear increase of the gain G with overvoltage. (b) Linear relation of the relative gain G_{rel} and the gain G.

In figure 3.6, V_{BD}^{on} and V_{BD}^{off} of the PM3350T STD are compared in the temperature range $-20\ ^\circ\text{C} \leq T \leq 30\ ^\circ\text{C}$. For both definitions of the breakdown voltage, a linear decrease is observed with temperature. The reason is, that the charge carrier mobility in silicon increases with decreasing temperature and consequently an equivalent ionization rate is achieved at lower voltages. At every temperature, V_{BD}^{off} is observed to be larger that V_{BD}^{on}. The difference of both quantities increases from 0.16 V at $T = 30\ ^\circ\text{C}$ to 0.3 V at $-20\ ^\circ\text{C}$. This observation is not understood, yet.

For most applications, an operation at stable photon detection efficiency and stable gain conditions is required. In figure 3.7, the relative gain is shown as a function of the overvoltage at different temperatures. No dependence of G_{rel} on T is observed. For this reason, V_{BD}^{off} is referred to as the breakdown voltage V_{BD} in the remainder of this work and is used to calculate the overvoltage $\Delta V = (V - V_{BD})$.

3. Determination of Standard SiPM Parameters

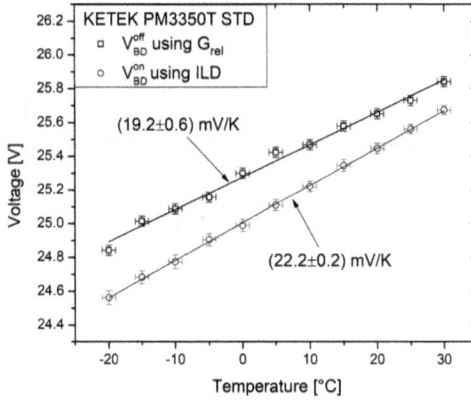

Figure 3.6: Breakdown voltage versus the temperature for the PM3350T STD. The breakdown voltages V_{BD}^{on} and V_{BD}^{off} are compared.

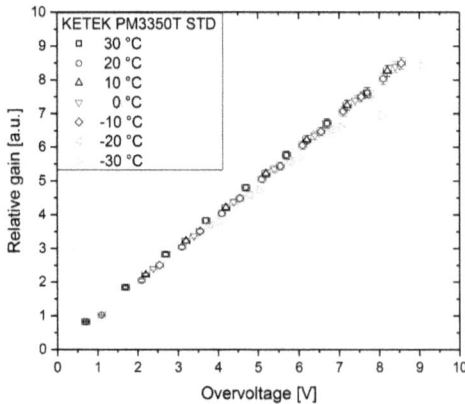

Figure 3.7: Relative gain of the PM3350T STD versus the overvoltage at different temperatures.

46

3.5 Photon detection efficiency

For the measurement of the photon detection efficiency, a commonly used procedure was applied [81], [82]. In this procedure, the SiPM is homogeneously illuminated by a pulsed laser (here, *PLP-10, Hamamatsu*, 406 nm, < 70 ps pulse length, 1 kHz repetition rate) with the optical output power P_{opt} and the pulse length τ. The SiPM response after each light pulse is recorded with an oscilloscope. In a signal trace analysis, the SiPM pulses are counted in a time-gate of 10 ns, which is centered at the expected arrival time of the photons. The number of detected photons per light pulse follows a Poisson distribution with the mean μ. An expression for μ is given in equation 3.7.

$$\mu = -\ln\left(\frac{N_0}{N_{tot}}\right) \tag{3.7}$$

Here, N_{tot} is the total number of light pulses and N_0 is the number of light pulses for which the SiPM did not detect a photon. This quantity is immune to optical crosstalk and afterpulsing [82].

The equation given above does not account for dark pulses which accidentally coincide with the laser pulse and hence lead to an artificially enhanced μ. To account for this effect, the measurement is repeated without actually illuminating the SiPM. The number of SiPM traces without a dark pulse in the 10 ns time-gate is given by N_0^{dark}.

The corrected mean number of detected photons per light pulse μ is then calculated as shown in equation 3.8.

$$\mu = -\ln\left(\frac{N_0}{N_{tot}}\right) + \ln\left(\frac{N_0^{dark}}{N_{tot}}\right) \tag{3.8}$$

The photon detection efficiency can be expressed as shown in equation 3.9. Here, h is the Planck constant, ν the frequency of the incident light, μ the number of photons and δ the fraction of the output power that reaches the SiPM.

$$PDE = \frac{\text{number of detected photons}}{\text{number of incident photons}} = \frac{\mu \cdot h\nu}{P_{opt} \cdot \tau \cdot \delta} \tag{3.9}$$

The main focus of this work is the understanding of the dark count rate of SiPMs. Consequently, the avalanche triggering probability P_{trigg} is a crucial parameter. The quantification of the absolute photon detection efficiency of the SiPMs is not required. The samples, which are directly compared in this work, have an equal geometric fill factor ε. Additionally, it is assumed that the quantum efficiency QE is equal for every sample, since the variation of the doping concentration is small. For this reason, the quantity μ is sufficient for a relative comparison of P_{trigg}.

In figure 3.8, μ is plotted versus the overvoltage for the PM3350T STD at $T = 30\ ^\circ\mathrm{C}$ and $T = -30\ ^\circ\mathrm{C}$. The dependence of μ on the overvoltage is in qualitative agreement with the results shown in figure 2.9(b) and indicates that the avalanche breakdowns are mostly triggered by electrons. No temperature dependence of μ is observed in this temperature range. The data points can be well described by the exponential function shown in equation 3.10. The parameter α depends on the geometry of the multiplication region, the impact ionization coefficients for electrons and holes as well as the shape of the electric field distribution. For this reason α is a wavelength and device specific parameter [82].

$$P_{trigg} \propto \mu = \mu_{max} \cdot \left(1 - \exp\left[-\frac{\Delta V}{\alpha}\right]\right) \tag{3.10}$$

As discussed in section 2.5, DCR is a function of P_{trigg}. From the obtained result, it is concluded that DCR values, which are measured in the temperature range $-30\ ^\circ\mathrm{C} \le T \le 30\ ^\circ\mathrm{C}$, can be directly compared without correcting for a varying P_{trigg}.

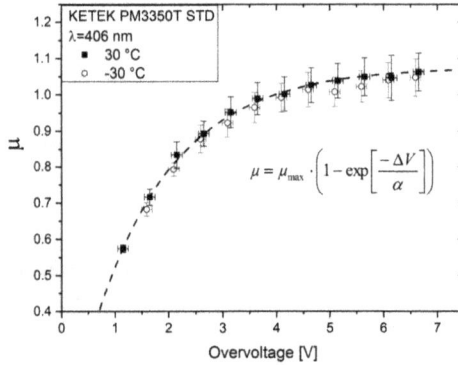

Figure 3.8: Mean number of detected photons versus the overvoltage at two distinct temperatures.

3.6 Dark count rate

The dark count rate was determined by the analysis of randomly triggered 1 μs long signal traces. It was calculated as the average number of pulses with an amplitude larger than 0.5 p.e. in one trace, divided by the length of the signal trace.

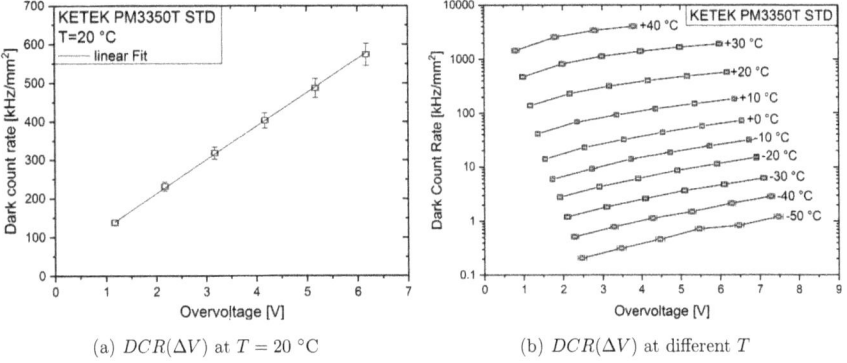

(a) $DCR(\Delta V)$ at $T = 20\ °C$ (b) $DCR(\Delta V)$ at different T

Figure 3.9: Dark count rate of the PM3350T STD. The lines in (b) are drawn as eye-guides.

As discussed in section 2.5, the dark count rate is described by a superposition of the charge carrier generation rates of different mechanisms multiplied by the avalanche triggering probability. The ratio of the contributing mechanisms depends on the internal structure of the SiPM micro-cells and changes with temperature. Consequently, the dependence of DCR on ΔV cannot be predicted in general for every SiPM type. For the KETEK SiPMs with a micro-cell size of 50x50 μm^2, $DCR(\Delta V)$ can be approximated by a linear function within the operation range of the devices, as shown in figure 3.9(a).

In figure 3.9(b), the dark count rate of the PM3350T STD is plotted versus the overvoltage for different temperatures. At $\Delta V = 4$, a decrease from approximately 4 MHz/mm^2 to approximately 0.4 kHz/mm^2 is observed, as the temperature decreases from 40 °C to -50 °C. This measurement demonstrates that cooling of the SiPM is an effective method for the suppression of dark pulses. From another point of view, a temperature stabilization has to be implemented for applications which require a SiPM operation at stable DCR conditions. As discussed in section 2.4, an enhancement of PDE can be achieved by increasing the micro-cell size. However, this will lead to an increase of the absolute dark count rate, since the active area will increase with an increasing fill factor. This was experimentally confirmed by comparing the dark count rate of KETEK PM11 SiPMs (chip dimensions: 1.2x1.2 μm^2) with different micro-cell sizes. The results are shown in figure 3.10(a). For the case that the dark count rate is normalized to the active area of the SiPM, no impact of the micro-cell size on DCR is observed (see figure 3.10(b)). This is valid at the condition that the dark count rate of one micro-cell is much smaller than the reciprocal recovery time. If the frequency of dark pulses of one micro-cell approaches the reciprocal recovery time, a larger fraction of

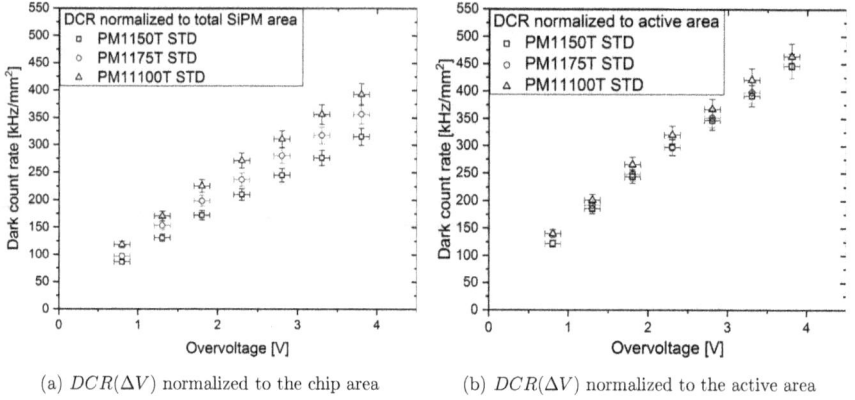

(a) $DCR(\Delta V)$ normalized to the chip area   (b) $DCR(\Delta V)$ normalized to the active area

Figure 3.10: Dark count rate versus the overvoltage for the PM11 (1.2x1.2 mm^2) with different micro-cell sizes.

dark pulses will occur at incomplete recovery and will be consequently not detected. This will artificially decrease the dark count rate of the SiPM but also its photon detection efficiency. The goal in SiPM development is to achieve the highest possible fill factor with the lowest possible noise. For this reason, the denoted dark count rates in this work are normalized to the total SiPM area.

3.7 Correlated noise

In contrast to dark pulses, which are independent of their history, correlated pulses are conditioned by their preceding events. In this work, two types of correlated pulses are distinguished:

(i) Afterpulsing, which is a delayed breakdown of the preceding firing micro-cell (see section 2.8).

(ii) Delayed crosstalk, which is a delayed breakdown of a neighbor of the preceding firing micro-cell (see section 2.7).

Both types of correlated pulses may reach amplitudes larger than 0.5 p.e. and consequently lead to an artificially enhanced DCR, if the method discussed in section 3.6 is applied. In this work, the probability of correlated pulses is determined by using the method proposed by Vinogradov [83]. The method is based on the analysis of the complementary cumulative distribution function (CCDF) of pulses subsequent to a primary dark pulse. The cumulative distribution function $P(t)$ is the probability that a positive event occurs at the time less than or equal to t. Alternatively, $P^*(t) = 1 - P(t)$ describes the probability that no positive event occurs at the time $\leq t$. Consequently, the probability that after a primary dark pulse no subsequent dark pulse occurs in the time interval Δt, is given by $P^*_{DCR}(\Delta t)$.

$$P^*_{DCR}(\Delta t) = \exp\left(-DCR \cdot \Delta t\right) \qquad (3.11)$$

The probability P^*_{tot} that neither a dark pulse nor a correlated pulse occurs in the time interval Δt after a primary dark pulse is given as the product of the independent probabilities $P^*_{DCR}(\Delta t)$ and $P^*_{Corr}(\Delta t)$.

$$P^*_{tot}(\Delta t) = \exp\left(-DCR \cdot \Delta t\right) \cdot P^*_{corr}(\Delta t) \qquad (3.12)$$

As mentioned in section 2.8, the time distribution of afterpulses is a superposition of the contributing trap levels. The additional contribution from delayed crosstalk pulses makes the prediction of the total time distribution of correlated events even more complicated. The advantage of the presented method is, that it allows for a reconstruction of the correlated pulse time distribution without any assumptions on its timing models [83] (see equation 3.13).

$$P_{corr}(\Delta t) = 1 - (P^*_{tot} \cdot \exp\left[DCR \cdot \Delta t\right]) \qquad (3.13)$$

However, in literature typically an exponential time distribution of correlated pulses with the probability P_{CP} and the characteristic time τ_{corr} is assumed. For this case, P^*_{corr} can be expressed as shown in equation 3.14.

$$P_{corr}(\Delta t) = P_{CP} \left(1 - \exp\left[-\frac{\Delta t}{\tau_{corr}} \right] \right) \qquad (3.14)$$

The determination of the probability of correlated pulses consists of five steps:

(i) Recording of signal traces:

A total number of N_{tot} SiPM signal traces with a length of 1 μs is recorded with the trigger threshold set at 0.5 p.e.. The trigger time is set such that the signal trace contains a time-gate of at least 100 ns before the SiPM pulse. In figure 3.11, an example of such a signal trace is shown. In this particular example, the primary pulse is followed by an afterpulse with an evidently reduced amplitude. The analysis of the data is performed offline.

(ii) Selection of valid signal traces:

A signal trace is considered valid, if it contains a SiPM pulse with an amplitude of 1 p.e. and no preceding pulses within 100 ns before the primary pulse. This restrictions are applied in order to ensure that the primary pulse is a single, independent dark pulse and not a correlated pulse itself. Signal traces which do not fulfill this requirement are excluded from the analysis.

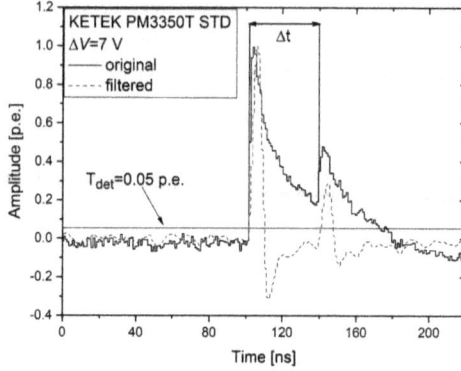

Figure 3.11: Example of a signal trace with a primary dark pulse and an afterpulse.

(iii) Determination of Δt:

After a signal trace with a primary dark pulse is considered valid, a possible subsequent pulse is searched and its delay-time Δt is determined with respect to the primary dark pulse. Only the first subsequent pulse is considered in the analysis. The threshold for the detection of subsequent pulses is set as low as possible under the condition of not detecting electronic noise (see figure 3.11). If no subsequent pulse is found, it is assumed that the primary pulse is followed by a subsequent dark pulse with Δt larger than the recorded signal trace.

(iv) Generation of the CCDF:

All determined delay-times Δt_i are sorted in a descending order. The probability $P^*_{tot}(\Delta t_i)$ is then attributed to the i-th delay-time as shown in equation 3.15. Here, N_0 is the number of traces without a subsequent pulse within the recorded time-gate and N_{tot} is the total number of valid traces. An example of the resulting $P^*_{tot}(\Delta t)$ is shown in figure 3.12.

$$P_{tot}(\Delta t_i) = \frac{N_0 + i}{N_{tot}} \tag{3.15}$$

(v) Extraction of DCR and P_{CP} from P^*_{tot}:

For sufficiently long delay-times, P^*_{corr} approaches the time independent probability $(1 - P_{CP})$ and equation 3.12 simplifies to

$$P^*_{tot}(\Delta t \gg 1) \approx (1 - P_{CP}) \cdot \exp\left(-DCR \cdot \Delta t\right) \tag{3.16}$$

At the condition that P_{corr}^* decays faster than P_{DCR}^*, the dark count rate and the probability of correlated pulses can be extracted by fitting equation 3.16 to the data points at large Δt. For the example given in figure 3.12, the CCDF is dominated by the time-distribution of dark pulses for $\Delta t \gtrsim 100$ ns. According to equation 3.15, $P_{tot}(\Delta t_{min}) = 1$ by definition. Here, Δt_{min} is the smallest detected delay-time. Since it is not possible to detect pulses at $\Delta t = 0$ by experimental approaches, the probability of correlated events is determined at Δt_{min} using the approximation which is shown in equation 3.17.

$$(1 - P_{CP}) \cdot \exp\left(-DCR \cdot \Delta t_{min}\right) \approx (1 - P_{CP}) \tag{3.17}$$

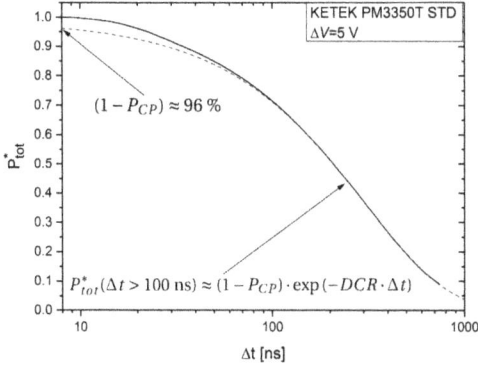

Figure 3.12: Example of a CCDF of subsequent pulses for the PM3350T STD.

In figure 3.13, the probability of correlated pulses is plotted versus the overvoltage for the PM3350T STD at $T = 21$ °C (filled squares). A $(\Delta V)^2$ dependence of P_{CP} is observed, which is in agreement with the results reported in [84]. The open circles in figure 3.13 show $P_{CP}(\Delta V)$ for $T_{det} = 0.5$ p.e.. The probability that a correlated pulse with an amplitude larger that 0.5 p.e. occurs after an initial dark pulse is less than 1 % for overvoltages up to 6 V. A variation of the ambient temperature by ≈ 1 °C from $T = 21$ °C, leads to a variation of DCR by approximately 13 %, for the PM3350T STD. Consequently, the contribution from afterpulses and delayed crosstalk pulses to the dark count rate is negligible, if determined with the method presented in section 3.6.

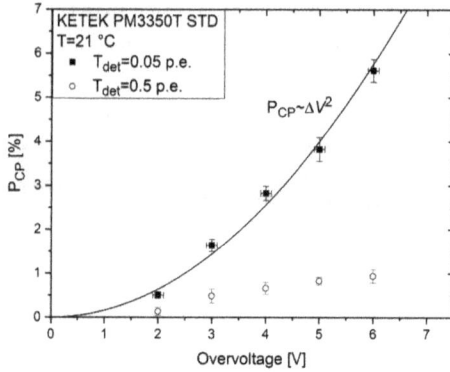

Figure 3.13: Probability of correlated pulses versus the overvoltage for the PM3350T STD. P_{CP} was determined for two distinct amplitude threshold levels T_{det}.

3.8 Prompt optical crosstalk probability

The prompt optical crosstalk probability P_{CT} is estimated as shown in equation 3.18. Here, DCR_n is the dark count rate measured with a discriminator threshold of n p.e. [1], [81] [85], [82]. In figure 3.14(a), the dark count rate of the KETEK PM3350T STD is shown as a function of the discriminator threshold. The corresponding pulse-height spectrum is shown in figure 3.14(b).

$$P_{CT} \approx \frac{DCR_{1.5}}{DCR_{0.5}} \tag{3.18}$$

Pulses with amplitudes > 1 p.e. are not exclusively generated by CT. Accidental coincidences of two or more dark events also lead to pulses with a larger amplitude. Since the occurrence of a dark pulse is independent of its history, the probability P_{DCR}, that a dark pulse occurs within the time t after a primary dark pulse, is given by equation 3.19.

$$P_{DCR} = 1 - \exp\left(-DCR_{0.5} \cdot t\right) \tag{3.19}$$

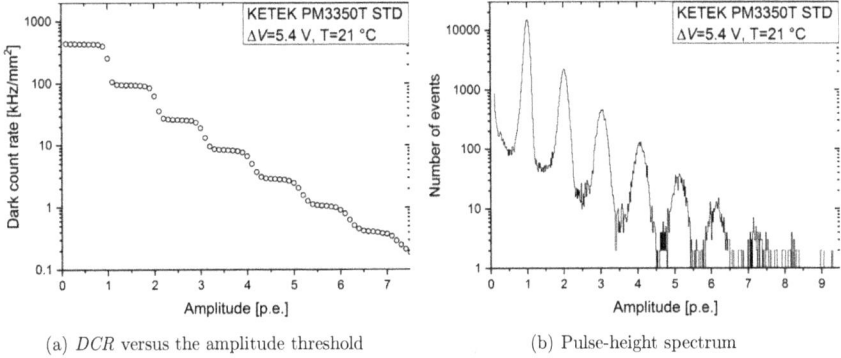

(a) DCR versus the amplitude threshold   (b) Pulse-height spectrum

Figure 3.14: (a) Dark count rate as a function of the discriminator threshold at $\Delta V = 5.4$ V and $T = 21$ °C. (b) Corresponding pulse-height spectrum.

The probability of not detecting a subsequent pulse after the occurrence of a primary dark pulse is a product of the probability of not detecting a CT pulse and the probability of not detecting a dark pulse. Consequently, the prompt optical crosstalk probability corrected for accidental coinciding dark pulses can be expressed as shown in equation 3.20 [86]. Here, Δt_{crit} is the time by which two consecutive pulses have to be separated in order to be distinguished. For dark count rates $DCR_{0.5} << 1/\Delta t_{crit}$, equation 3.20 converges to equation 3.18.

$$P_{CT} = 1 - \left[1 - \frac{DCR_{1.5}}{DCR_{0.5}} \right] \cdot \exp\left(DCR_{0.5} \cdot \Delta t_{crit} \right) \qquad (3.20)$$

In figure 3.15, the prompt optical crosstalk probability of the KETEK PM3350T STD is shown as a function of overvoltage, measured at different temperatures. For lower overvoltages P_{CT} shows a ΔV^2 dependence, which is in agreement with the observations reported in [87]. This can be explained by the fact that P_{CT} is proportional to $PDE(\Delta V)$ and $G(\Delta V)$. Once the saturation of PDE is reached at high enough overvoltages, P_{CT} starts to show a linear dependence on ΔV.

No temperature dependence of P_{CT} is observed, which is in agreement with the results reported in [88]. This observation is crucial for the experimental methods presented in this work and indicates that neither the gain of the SiPM nor the photon detection efficiency change with temperature.

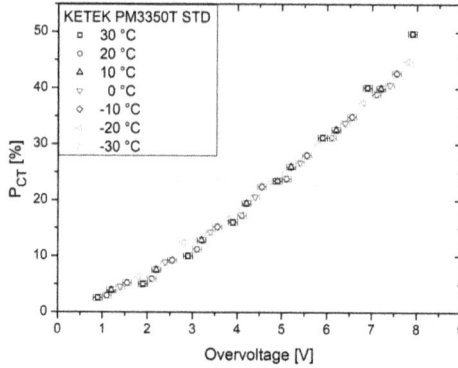

Figure 3.15: Probability of prompt optical crosstalk P_{CT} versus the overvoltage at different temperatures.

3.9 Investigated samples

The SiPM types which were used for the investigation of the dark count rate are listed in table 3.2. The key parameters which are necessary for the line of argument in this work are indicated. Here, E_{Imp}^P is the energy of phosphorus used for the implantation of the n-layer and "Impl. dose" is the implanted phosphorus dose. As will be pointed out in chapter 7, the parameters of the n-layer implantation are of major importance for the suppression of the dark count rate of the KETEK SiPMs. V_{BD} is the breakdown voltage, N_{cells} is the total number of micro-cells and "Pitch" describes the edge length of one micro-cell.

Two versions of the PM3350T are characterized in this work. The outer dimensions of version 1 are shown in figure 3.16(a). The layout of the corresponding micro-cell is shown in figure 3.16(d). Version 2 of the PM3350T has modified outer dimensions and a slightly reduced number of micro-cells, which is shown in figure 3.17(a). Figure 3.17(c) shows the micro-cell layout of version 2. The n-layer and hence the fill factor of version 2 is increased with respect to version 1. The outer mico-cell dimensions of version 1 and version 2 do not differ. In the remainder of this thesis, the two versions of the PM3350T are not explicitly distinguished. If two technologies are directly compared, the same versions are used, unless stated otherwise. The specified variation of the breakdown voltage for the PM3350T is not connected with the different versions but with the fab in which the devices were fabricated. The PM4450T has a significantly increased number of micro-cells with respect to the PM3350T.

Consequently its outer dimensions are larger, as shown in figure 3.17(b). The micro-cell layout is the same as for the version 2 of the PM3350T (see figure 3.17(c)). This SiPM type is only used for the characterizations discussed in section 7.3.

Table 3.1: List of investigated samples - Dark count rate

Series	PM3350T STD	PM3350T HE	PM3350T HE-S	PM4450T STD	PM4450T LD
E_{Imp}^P [MeV]	3.5	4.5	4.5	3.5	3.5
Impl. dose [cm^{-2}]	$1 \cdot 10^{13}$	$1 \cdot 10^{13}$	$1 \cdot 10^{13}$	$1 \cdot 10^{13}$	$5 \cdot 10^{12}$
V_{BD} [V]	$25.0 - 27.1$	$34.6 - 35.4$	26.2 ± 0.1	27.1 ± 0.1	32.6 ± 0.1
N_{cells}	$3568-3600$	$3568-3600$	3568	5056	5056
Pitch [μm]	50	50	50	50	50

In chapter 8, the degradation of the SiPM parameters after the irradiation with ^{60}Co γ-rays and with thermal neutrons is discussed. For these investigations, SiPMs with a reduced number of micro-cells were chosen, in order to be able to handle the high absolute dark count rate after the irradiation. The used samples are listed in table 3.2. The micro-cell structure of both SiPMs is the same as for the version 1 of the PM3350T.

Table 3.2: List of investigated samples - Radiation hardness

Series	PM1150T STD	6x6 test-structure
E_{Imp}^P [MeV]	3.5	3.5
Impl. dose [cm^{-2}]	$1 \cdot 10^{13}$	$1 \cdot 10^{13}$
V_{BD} [V]	26.0 ± 0.1	27.4 ± 0.1
N_{cells}	576	36
Pitch [μm]	50	50
Radiation type	^{60}Co γ-rays	thermal neutrons

(a) PM3350T version 1

(b) PM1150T

(c) 6x6 test-structure

(d) Micro-cell layout version 1

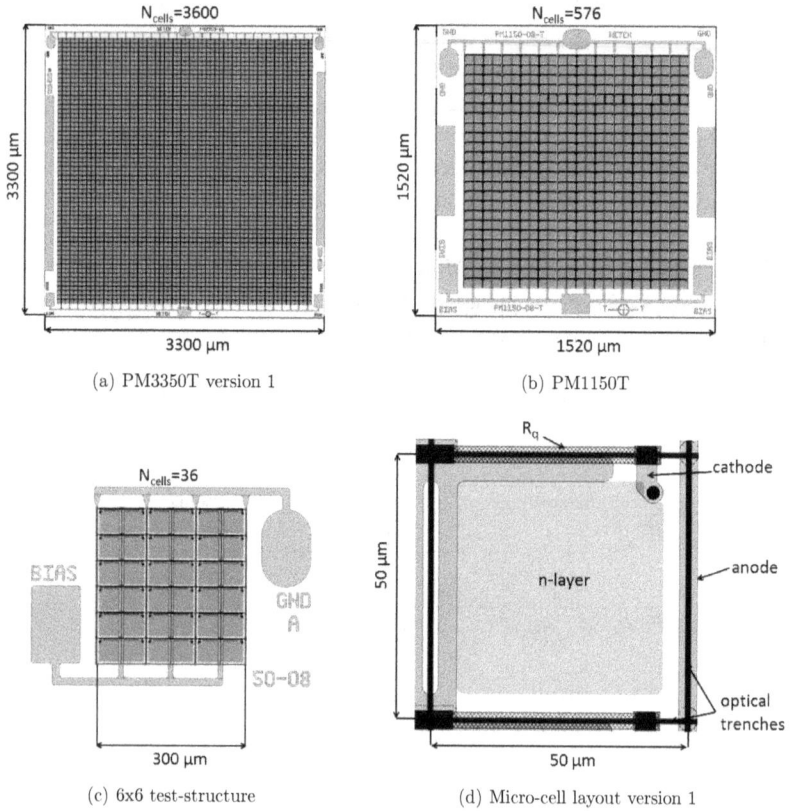

Figure 3.16: Chip and micro-cell layouts.

(a) PM3350T version 2

(b) PM4450T

(c) Micro-cell layout version 2

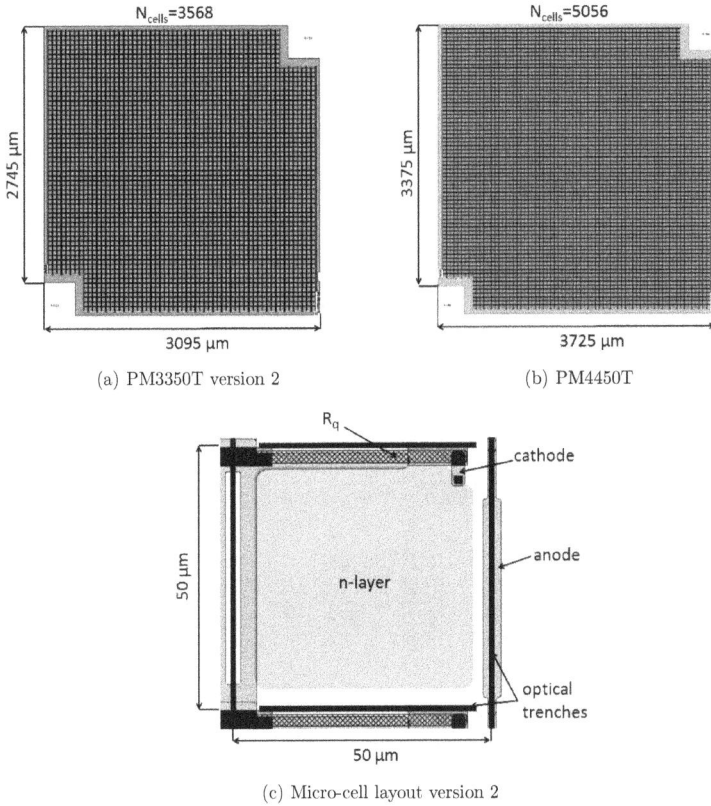

Figure 3.17: Chip and micro-cell layouts.

Chapter 4

Identification of Mechanisms Responsible for Dark Pulses

In section 2.5, the potential mechanisms responsible for the occurrence of dark pulses were discussed. One central goal of this work is to identify the physical mechanisms which lead to an enhanced dark count rate of the KETEK PM3350T STD. In this chapter, two approaches to fulfill this task are presented. The main idea of both is to extract characteristic activation energies E_{act} of the contributing mechanisms from the temperature dependence of the dark current I_{dark}.

In section 4.1, the utilized experimental setup is presented. A widely reported conventional approach is discussed in section 4.2. In section 4.3, a novel approach is presented, which was developed within the scope of this work.

4.1 Experimental setup

In order to investigate the SiPMs at different temperatures, the devices were mounted onto a thermo-electric cooler (TEC) and evacuated in a TO-8 housing with a quartz window. An example is shown in figure 4.1. Stable temperatures from $+100\,^\circ$C down to $-35\,^\circ$C could be reached with this setup. The schematic of the complete experimental setup is shown in figure 4.2. The measurements were performed in a dark box. Because the method presented in section 4.3 requires an illumination of the SiPM, the output of a tunable light source (*Optometrics, TLS-6* [89]) was coupled to an optical fiber and the fiber output was placed in front of the SiPM. It was assured that the total active area was illuminated uniformly. The absorption coefficient in silicon for $\lambda = 600$ nm at 298 K is $\alpha \approx 4.5 \cdot 10^3$ cm^{-1} [90]. This wavelength was chosen for the experiments in order to probe the micro-cells to a depth larger than 2 μm. The peak of the phosphorus concentration forming the n-layer is located at roughly 1.6 μm within the silicon.

The current-voltage characteristics were measured with a *Keithley 4200 Parameter Analyzer*. For the temperature readout, a Pt100 temperature sensor was placed close to the SiPM within the TO-8 housing, as indicated in figure 4.1. For the characterization of the SiPM in pulsed mode, the *DRS4 Evaluation Board*, provided by *Paul Scherrer Institut*, was used. The data was acquired with a bandwidth of 700 MHz and a sampling rate of 1GS/s. In order to investigate SiPMs which were not mounted onto a TEC, a modified experimental setup was used. The SiPM and the output of the optical fiber were placed in a light-tight, calibrated *Vötsch, VCL 7003* climate chamber. In addition to the build-in temperature sensor of the chamber, the temperature in the close vicinity of the SiPM was monitored via an additional Pt100 sensor.

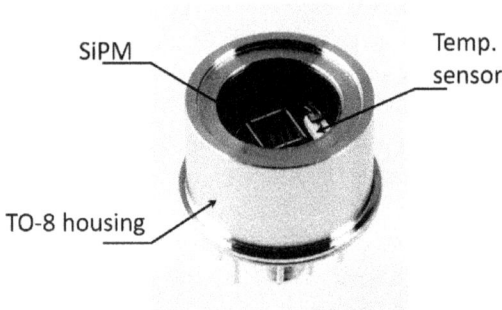

Figure 4.1: Example of a KETEK PM3350T SiPM which is mounted onto a TEC and evacuated in a TO-8 housing with a quartz window.

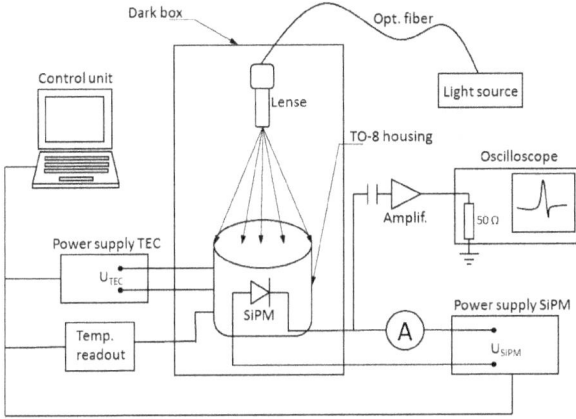

Figure 4.2: Sketch of the experimental setup for the characterization of SiPMs at adjustable temperatures.

4.2 Conventional method

The conventional method for the determination of the activation energy of contributing mechanisms is widely reported in literature, as for example in [48] [91], [92] and [93]. The method is based on the Arrhenius plot. The natural logarithm of the measured quantity (here, I_{dark}) is plotted as a function of $(1/kT)$, with k being the Boltzmann constant and T being the temperature. The activation energy E_{act} is determined by fitting equation 4.1 to the experimental data. Here, $\gamma = 3.2$ for $T \geq 10$ °C and $\gamma = 2.1$ for $T \leq -10$ °C was used, according to equations 2.37 and 2.42. E_{act} and A are free parameters.

$$\ln(I_{dark}) = A + \ln(T^\gamma) - \frac{E_{act}}{kT} \tag{4.1}$$

In [48], a correlation of I_{dark} and DCR was reported in the temperature range from -25 °C to $+65$ °C. This correlation was confirmed for the KETEK PM3350T STD SiPM in the temperature range from $+20$ °C to -20 °C, as shown in figure 4.3. The lines in figure 4.3 show the measured I_{dark}. The open symbols represent the reconstructed I_{dark} from the measured DCR, using equation 4.2. Here, q is the elementary charge, G is the gain and η is a quantity accounting for prompt optical crosstalk. The good agreement of the experimental data with the model, indicates a low contribution of afterpulses to the dark current.

$$I_{dark}(V \geq V_{BD}) = q \cdot G \cdot DCR \cdot \eta \tag{4.2}$$

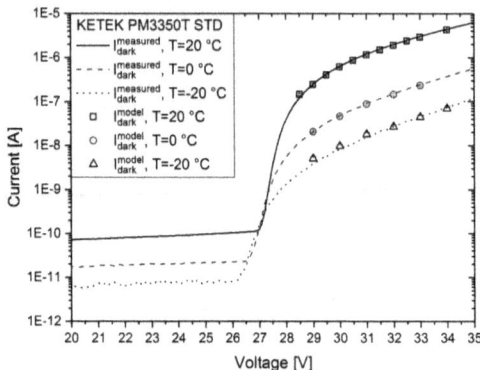

Figure 4.3: Correlation of the experimentally determined I_{dark} (lines) and I_{dark} reconstructed from DCR (open symbols).

In the following, the quantity η is explained in more detail. As discussed in section 2.7, a prompt optical crosstalk pulse occurs simultaneously with the primary dark pulse. Consequently, a combination of a primary dark pulse which is followed by n prompt optical crosstalk pulses is counted as one pulse. However, the charge output of such a pulse is $(n+1)$ times larger with respect to a single dark pulse. For this reason, the measured DCR must be corrected for the prompt optical crosstalk probability P_{CT} in order to correlate the dark current with the dark count rate. Vinogradov [94] analytically modeled the optical crosstalk as a branching Poisson process. According to this model, every pulse produces a Poisson distributed number of crosstalk pulses with the mean ν. Equation 4.3 gives the relation between ν and P_{CT}. The total number of pulses in the process containing the single primary dark pulse and all crosstalk pulses obeys a Borel distribution with the mean η, given in equation 4.4.

$$\nu = -\ln\left(1 - P_{CT}\right) \tag{4.3}$$

$$\eta = \frac{1}{1 - \nu} \tag{4.4}$$

In chapter 3, it was shown that G and P_{CP} are temperature independent for a fixed overvoltage (see figures 3.7 and 3.15). From equation 4.2, it is concluded that the temperature dependence of the dark current is only determined by the temperature dependence of the dark count rate. Consequently, the activation energies determined from the Arrhenius plots of I_{dark} are also valid for the dark count rate.

64

4.2.1 Results

In this section, the dark current of the PM3350T STD is characterized with the help of the conventional method. The characterization was performed in two temperature ranges with an increment of 5 °C. The high temperature range extends from $T = 30$ °C to $T = 10$ °C and the low temperature range extends from $T = -10$ °C to $T = -30$ °C. In the following, the obtained E_{act} at fixed voltage and fixed overvoltage conditions are discussed.

Figure 4.4(a) shows the Arrhenius plots of I_{dark} at four different voltages. A faster decrease of I_{dark} with T in the high temperature range with respect to the low temperature range is clearly evident. The corresponding activation energies are shown in figure 4.4(b). In the high temperature range, E_{act} increases from (0.67 ± 0.05) eV at 28 V to (0.95 ± 0.02) eV at 35 V. In the low temperature range, E_{act} increases from (0.30 ± 0.01) eV at 28 V to (0.38 ± 0.01) eV at 35 V. As discussed in section 3.4, a linear increase of V_{BD} with T is observed with a slope of approximately $dV/dT \approx 20$ mV/°C. Accordingly, ΔV decreases with increasing T for the case of a fixed operation voltage. A modification of ΔV has an influence on P_{CT}, P_{CP}, on the gain of the SiPM and on the Geiger discharge development. This leads to a modification of the SiPM dark response and a systematic underestimation of E_{act} when determined at a fixed voltage.

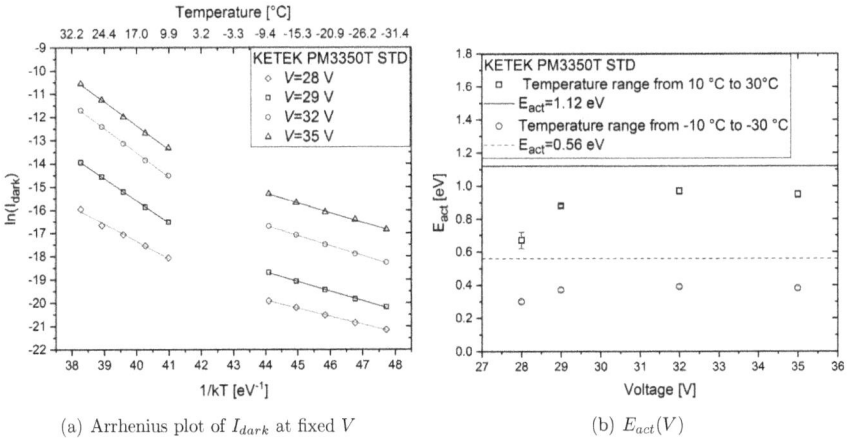

(a) Arrhenius plot of I_{dark} at fixed V (b) $E_{act}(V)$

Figure 4.4: Conventional determination of activation energies at fixed voltage conditions of the KETEK PM3350T STD.

In order to characterize the SiPM at fixed P_{CT}, P_{CP}, G and P_{trigg}, the analysis was performed at fixed ΔV. Figure 4.5(a) shows the Arrhenius plots of I_{dark} at 5 different overvoltages. The corresponding activation energies are displayed in figure 4.5(b). For the high temperature range, E_{act} is monotonically decreasing from (1.11 ± 0.05) eV at $\Delta V = 1$ V to (1.01 ± 0.03) eV at $\Delta V = 8$ V. In the low temperature range, E_{act} decreases from (0.50 ± 0.01) eV at $\Delta V = 1$ V to (0.45 ± 0.01) eV at $\Delta V = 8$ V. The observed decrease is qualitatively explained by the increasing contribution of electric field effects with ΔV as shown in figure 2.11.

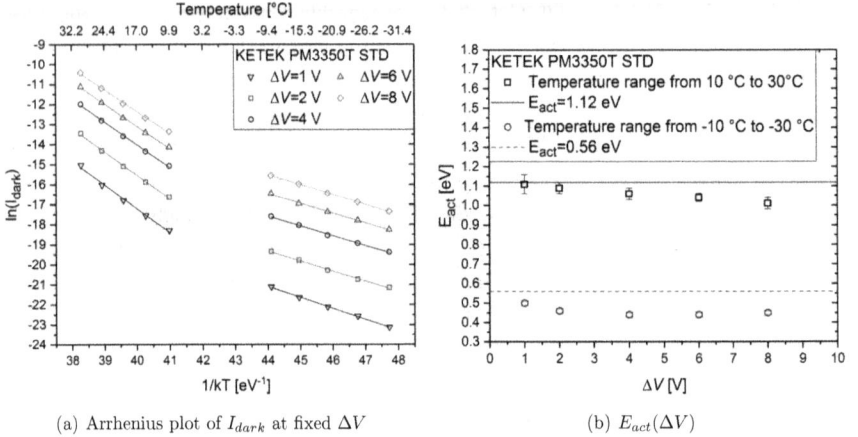

(a) Arrhenius plot of I_{dark} at fixed ΔV

(b) $E_{act}(\Delta V)$

Figure 4.5: Conventional determination of activation energies at fixed overvoltage conditions of the KETEK PM3350T STD.

At low field conditions, E_{act} determined in the high temperature region is close to the silicon energy bandgap of 1.12 eV. This result indicates that a dominant contribution from a diffusion current I_{diff} is present for temperatures $T \geq 10$ °C.

In the low temperature range, E_{act} is close to 0.56 eV at $\Delta V = 1$ V. This result implies a dominant contribution from generation-recombination centers with an ionization energy close to Si midgap by the process of SRH-Generation (see equation 2.42). The increasing deviation of E_{act} from 0.56 eV with increasing ΔV is anticipated to be caused by a non-negligible contribution of electric field effects to I_{dark} (see section 2.5.2).

The drawback of the conventional method is that the extracted E_{act} is determined by a superposition of all contributing mechanisms. The identification of a single mechanism is only possible if the contribution of all other mechanisms is negligible in a certain temperature range and for certain voltages/overvoltages. However, I_{dark} depends on a variety of non-negligible contributions. Some of them show a dependency on the applied voltage and some of them depend on the applied overvoltage.

Since the breakdown voltage of a GM-APD decreases with decreasing temperature, it is not possible to perform temperature dependent measurements at simultaneously fixed V and ΔV. For this reason, the conventional method is only suited for an estimation of E_{act} in order to identify the temperature regions in which certain contributions are dominating. However, for a more extended analysis of contributing mechanisms, a method is required which provides a separation of voltage dependent field effects and overvoltage dependent SiPM parameters. For this purpose, a novel characterization method was developed and is presented in the following section.

4.3 Novel method

The novel method is based on two independent measurements of the SiPM currents. In the first measurement, the photo-current is determined at illuminated SiPM conditions (here at $\lambda = 600$ nm). This provides access to the responsivity of the detector, which is an appropriate reference for I_{dark}. In the second measurement, the dark current is determined. This approach allows for the extraction of a field-independent expression of the initial amount of charge carriers I_{ini}, which is provided to or generated inside the multiplication region. From the Arrhenius plot of I_{ini}, characteristic activation energies were extracted, in order to identify the dominating contributions.

4.3.1 First approach

In the first step of the method, the responsivity R of the SiPM is defined for every temperature as expressed in equation 4.5. Here, I_{ph} is the photo-current of the SiPM, measured under illuminated conditions. The reference photo-current $I_{ph}(V_0)$ was determined at $V_0 = 5$ V. At this voltage, the gain of the APD is anticipated to be close to one. R accounts for the gain G and the avalanche initiation probability P_{trigg} of the SiPM, as well as for the optical crosstalk and afterpulsing contributions. The presented analysis is based on the assumption that the difference of the responsivity to photon-induced charge carriers R_{ph} and charge carriers originating from dark generation R_{dark}, is negligible. In figure 4.6, I_{dark}, I_{ph} and R_{ph} of the KETEK PM3350T STD are shown at $T = 20$ °C.

$$R_{dark}(\Delta V) \approx R_{ph}(\Delta V) = \frac{I_{ph}(\Delta)}{I_{ph}(V_0)} \qquad (4.5)$$

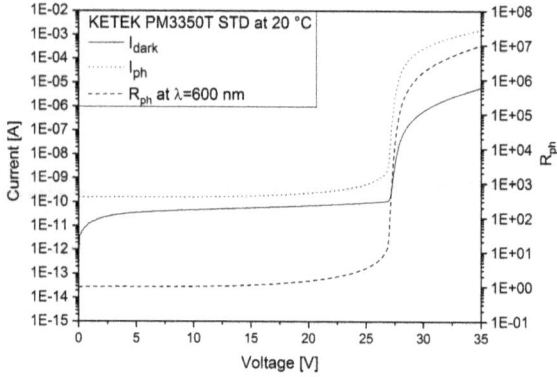

Figure 4.6: Voltage dependence of the dark current I_{dark}, the photo-current I_{ph} and the resulting responsivity R_{ph} of the KETEK PM3350T STD at $T = 20$ °C.

The dark current of the SiPM is divided into two contributions. The first contribution, I_{not_gained}, is associated with surface currents or peripheral currents which flow at low field conditions and dominate I_{dark} at $V < V_{BD}$. The major part of this contribution does not reach the high-field region of the GM-APD and thus is not multiplied. Using the conventional approach, the activation energy of I_{not_gained} was determined at different bias voltages $V < V_{BD}$ in the temperature range between +35 °C and +10 °C. The Arrhenius plots are shown in figure 4.7(a). The activation energies are close to $E_{gap}/2$ and decrease with increasing bias voltage, as shown in figure 4.7(b). This result leads to the conclusion that the dark current before breakdown is dominated by the mechanisms of SRH-Generation. I_{not_gained} is expected to increase with the increasing depletion width and hence was modeled with a square root dependence on the applied voltage [52], as expressed in equation 4.6. In figure 4.8, the agreement of the modeled I_{not_gained} with the experimentally determined $I_{dark}(V < V_{BD})$ is demonstrated for three temperatures.

$$I_{not\_gained} = I_{offset} + I_0 \cdot \left(\frac{V}{V_{eff}}\right)^{\frac{1}{2}}$$ (4.6)

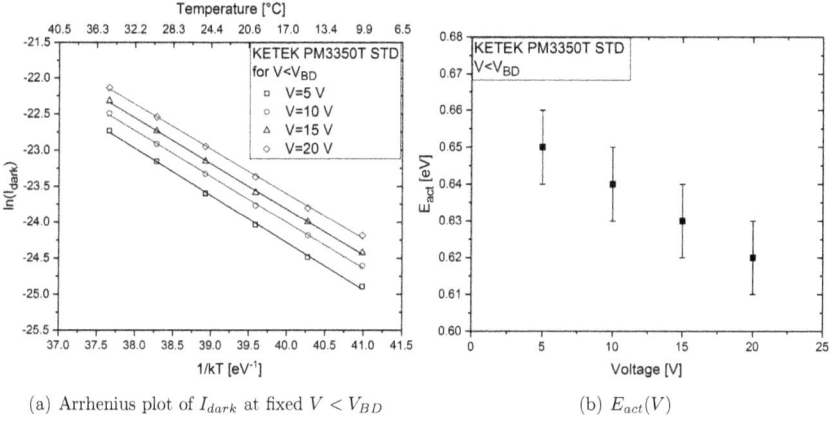

(a) Arrhenius plot of I_{dark} at fixed $V < V_{BD}$ (b) $E_{act}(V)$

Figure 4.7: Conventional determination of activation energies of $I_{dark}(V < V_{BD})$.

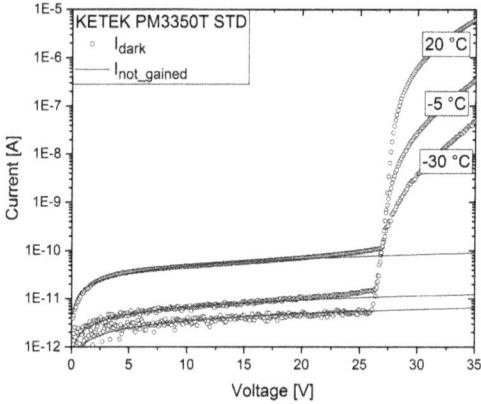

Figure 4.8: Comparison of measured and modeled surface and peripheral currents (I_{not_gained}) at different temperatures.

4. Identification of Mechanisms Responsible for Dark Pulses

The second contribution, I_{ini}, describes the initial amount of charge carriers which are generated inside the high-field region or are provided to this region via diffusion or drift. Since this part of the dark current reaches the multiplication region, it is affected by the responsivity R_{ph}. In equation 4.7, the initial expression for the modeling of I_{dark} is shown.

$$I_{dark}(V, \Delta V, T) = I_{not\_gained}(V, T) + I_{ini}(V, T) \cdot R_{ph}(\Delta V) \qquad (4.7)$$

In the first approach, I_{ini} is considered as a fraction δ of I_{not_gained} which does reach the multiplication region of the GM-APD. Under this assumption, equation 4.7 can be rewritten to equation 4.8.

$$I_{dark} = (1 - \delta) \cdot I_{not\_gained} + \delta \cdot I_{not\_gained} \cdot R_{ph} \qquad (4.8)$$

In figure 4.9(a), the dark current before multiplication (I_{dark}/R_{ph}), I_{not_gained} and I_{ini} are shown. Here, δ was chosen such that (I_{dark}/R) and I_{ini} coincide at the point of the avalanche breakdown.

In figure 4.9(b), the modeled dark currents are shown for three temperatures. In the first approach, contributions of electric field effects, as discussed in section 2.5, were not taken into account. The evident underestimation of the experimental data leads to the conclusion that electric field effects are significantly contributing to the dark current of the SiPM and have to be implemented into the model.

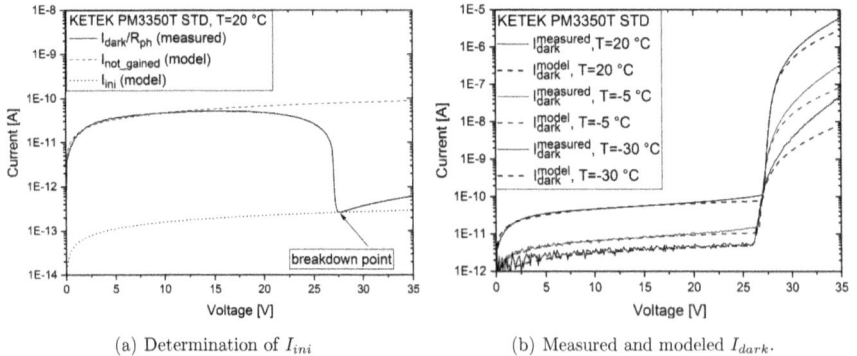

(a) Determination of I_{ini}

(b) Measured and modeled I_{dark}.

Figure 4.9: Model of I_{dark} of the PM3350T STD at different temperatures, resulting from the first approach.

4.3.2 Second approach

In the second approach, the model for I_{dark} is extended by the parameter F_{field}. The additional parameter accounts for those contributions to I_{dark}, which show an explicit dependence on the electric field. The expression for I_{dark} was consequently modified as shown in equation 4.9.

$$I_{dark}(V, \Delta V, T) = (1 - \delta) \cdot I_{not\_gained}(V, T) + I_{ini}(T) \cdot R_{ph}(\Delta V) \cdot F_{field}(V) \qquad (4.9)$$

In comparison to equation 4.7, I_{ini} in the second approach is modeled to be independent of the applied voltage and consequently independent of the electric field. The made approximation is justified in the following:

I_{ini} should account for contributions to I_{dark} which show no dependence or an implicit dependence on the electric field. Considering the mechanisms discussed in section 2.5, only the contribution from the pure SRH-Generation I_{gen} (see equation 2.41) and the contribution from the diffusion current $I_{diffusion}$ (see equation 2.33) are qualified to contribute to I_{ini}. The electric field dependence of the pure SRH-Generation is expressed by the dependence on the depletion region width $W(V)$.

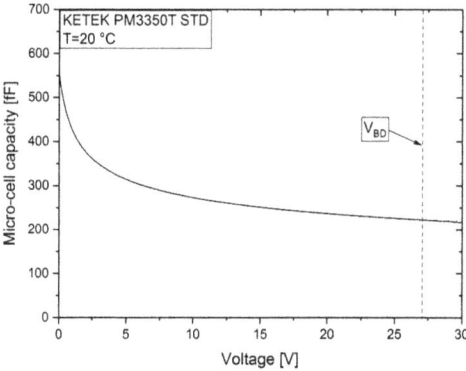

Figure 4.10: Micro-cell capacity of the PM3350T STD as a function of the applied voltage.

In figure 4.10, the measured micro-cell capacity C_{cell} is shown as a function of the applied voltage for the KETEK PM3350T STD. In the voltage range from $V = V_{BD}$ to $V = 30$ V, C_{cell} shows a decrease of about 2.3 %. Due to the weak decrease of C_{cell} with V in the region where $R_{ph} \gg 1$ is valid, the depletion width is anticipated to show a negligible increase with the applied voltage above breakdown. Consequently, the voltage dependence of I_{gen} is neglected.

The diffusion current $I_{diffusion}$ originates from a gradient of the charge carrier concentration as shown in equation 2.19. For the boundary condition of $V > V_{BD}$, $\nabla(n,p)$ is anticipated to show a slow variation with the applied voltage, analogously to W. Consequently, the increase of $I_{diffusion}$ with the bias voltage is approximated to be negligible.

The contribution to I_{dark} which does not pass the multiplication region and is consequently not affected by R_{ph}, is modeled as described in the first approach. The difference of the experimentally determined I_{dark} and the modeled $(1-\delta)\cdot I_{not\_gained}$ results in the multiplied current I_{diff}, which is expressed in equation 4.10. I_{diff} is approximated by a second order polynomial function of R_{ph}.

$$I_{diff} = I_{ini} \cdot R_{ph} \cdot F_{field} \approx I_{ini} \cdot R_{ph} \left(1 + \frac{R_{ph}}{R_{eff}} \right) \tag{4.10}$$

In figure 4.11(a), the experimentally determined and the modeled I_{diff} are shown at three temperatures. Only the data points for $R_{ph} \leq 1 \cdot 10^7$ were considered, which corresponds to $\Delta V \lesssim 7.8$ V. For $\Delta V \gtrsim 7.8$ V, the KETEK PM3350T STD reaches an overvoltage range in which a strong increase of I_{dark} is observed and the model is no longer applicable. In figure 6.14, the different overvoltage ranges are explained in more detail.

The developed model is in a better agreement with the experimental data in the lower temperature range. As shown in figure 4.11(b), the relative residuals at $T = -5\,^\circ$C and $T = -30\,^\circ$C are below 10 % in the whole responsivity range. For $T = 20\,^\circ$C, the model shows a deviation from the experimental data of less than 10 % for $R_{ph} < 7 \cdot 10^6$, which corresponds to $\Delta V = 6.5$ V. In figure 4.12, the measured and modeled dark currents are shown at three different temperatures. For voltages above the breakdown voltage, the model is in a good agreement with the experimental data within the uncertainties discussed above. The model shows a clear deviation from the measured currents several volts before the breakdown. This observation is not understood, yet. Since the main focus of this work is on the operation of the SiPM above the breakdown voltage, this effect was not investigated in the scope of this research project.

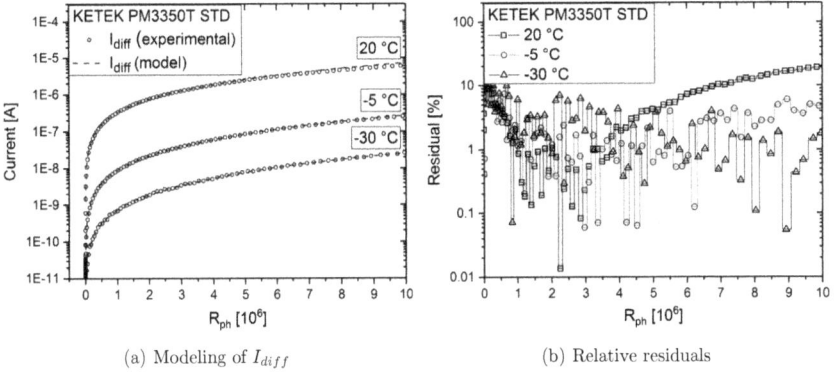

(a) Modeling of I_{diff}

(b) Relative residuals

Figure 4.11: Modeling of I_{diff} as a second order polynomial function of R_{ph}.

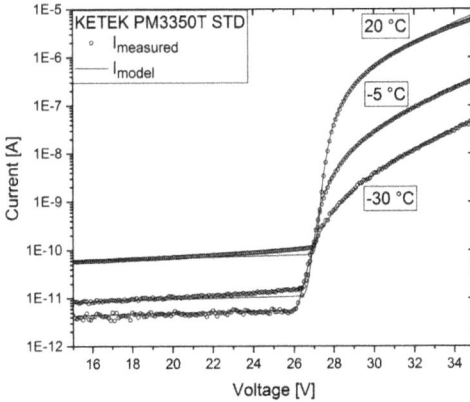

Figure 4.12: Model of I_{dark} of the PM3350T STD at different temperatures, resulting from the second approach.

73

4.3.3 Results

In figure 4.13, the Arrhenius plot of the extracted parameter I_{ini} is shown. The fit function 4.1 was used for the extraction of the activation energy, analogous to the conventional method. For $T \geq 10$ °C, an activation energy of (1.12 ± 0.01) eV $= E_g(Si)$ is observed which is in agreement with E_{act} determined with the conventional approach at $\Delta V = 1$ V. This result confirms that I_{dark} of the KETEK PM3350T STD is dominated by a diffusion of charge carriers at higher temperatures. For $T \leq -10$ °C, the electric field independent I_{ini} shows an activation energy of $E_{act} = (0.55 \pm 0.02)$ eV $\approx \frac{E_g(Si)}{2}$. This result identifies the SRH-Generation to be the dominant contributor to I_{dark} at lower temperatures.

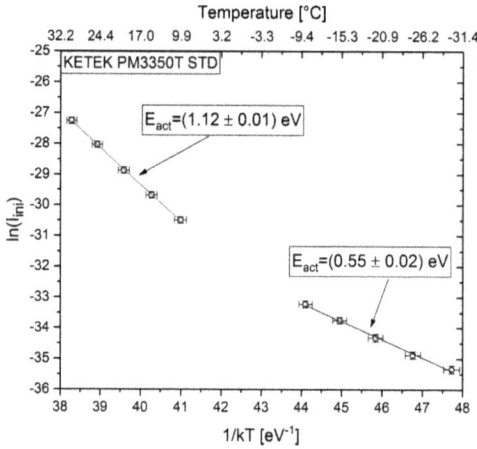

Figure 4.13: Arrhenius plot of I_{ini} of the PM3350T STD, determined with the second approach of the novel method.

Comparing equation 4.10 with equation 2.46, the expression $(1 + R_{ph}/R_{eff})$ is identified with the field-effect function $(1 + \Gamma)$. This is equivalent to the identification of the contributing field effects with the effect of trap-assisted tunneling. Using the equation 2.44 and 2.45, the temperature dependence of Γ can be simplified to

$$\Gamma \sim (kT)^{-\frac{3}{2}} \cdot \exp\left(\frac{q^2 \hbar^2 F_{eff}^2}{24 m_t (kT)^3}\right) \tag{4.11}$$

In order to estimate the effective electric field F_{eff}, the quantity $\ln(R_{ph}/R_{eff})$ was plotted as a function of $(1/kT)^3$ at a fixed voltage (see figure 4.14(a)). Equation 4.12 was fitted to the data points, with A and B being two free parameters. F_{eff} was then extracted from the parameter B using equation 4.13.

$$\ln(\Gamma) = A + \ln((kT)^{-\frac{3}{2}}) + B \cdot (kT)^{-3} \tag{4.12}$$

$$F_{eff} = \frac{24 m_t \cdot B}{q^2 \hbar^2} \tag{4.13}$$

Figure 4.14(b) shows F_{eff} as a function of the applied voltage. An average value of $F_{eff} \approx 5.5 \cdot 10^5$ Vcm^{-1} was determined. This is in agreement with the expected electric fields in the multiplication region of a GM-APD at operation conditions. Based on the predictions from figure 2.11, an effective electric field of $2.0 \cdot 10^5$ Vcm$^{-1} \lesssim F_{eff} \lesssim 3.5 \cdot 10^5$ Vcm^{-1} was expected. The determined values exceed these expectation by roughly a factor of 2. Additionally, a decrease from $F_{eff}(29\ \text{V}) = 5.9 \cdot 10^5$ Vcm^{-1} to $F_{eff}(35\ \text{V}) = 5.0 \cdot 10^5$ Vcm^{-1} is evident. This observation is interpreted as the existing non-negligible contribution from band-to-band tunneling J_{bbt}, as discussed in section 2.5.4. J_{bbt} increases with the applied voltage and does not depend on temperature (see equation 2.48). This leads to a reduction of the parameter B for higher voltages and consequently leads to a lowering of the extracted F_{eff}.

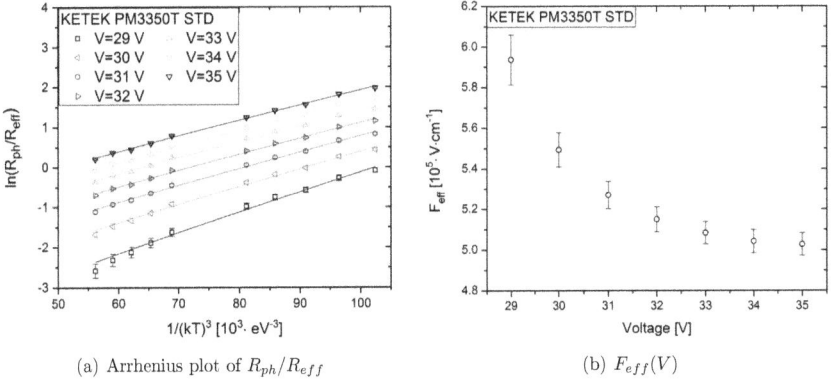

(a) Arrhenius plot of R_{ph}/R_{eff}

(b) $F_{eff}(V)$

Figure 4.14: Determination of the effective electric field F_{eff} of the PM3350T STD.

Figure 4.15 shows the contribution ρ of dark currents caused by electric field effects to the overall I_{dark}. As expected, ρ increases with increasing overvoltage and decreasing temperature. Due to this effect, E_{act} of I_{dark} determined with the conventional method, under the condition of low electric fields converges towards E_{act} of I_{ini}, determined with the novel method.

$$\rho = \frac{I_{ini} \cdot R_{ph} \cdot \frac{R_{ph}}{R_{eff}}}{I_{ini} \cdot R_{ph} \cdot (1 + \frac{R_{ph}}{R_{eff}})} = \frac{R_{ph}}{R_{eff} + R_{ph}} \qquad (4.14)$$

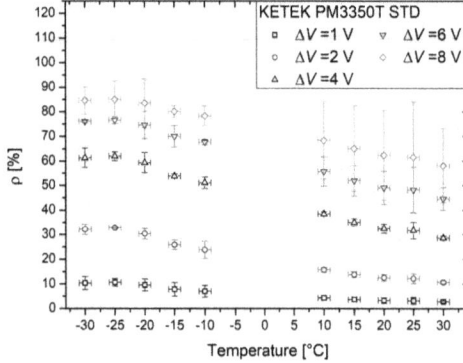

Figure 4.15: Contribution of dark currents due to electric field effects to the overall I_{dark}.

4.4 Conclusion

In this chapter, the dark current of the KETEK PM3350T STD was analyzed as a function of temperature and the contributing physical mechanisms were identified. Using the conventional approach under the condition of a low electric field, charge carrier diffusion was identified to determine I_{dark} at $T \geq 10$ °C. For temperatures $T \leq -10$ °C, the SRH-Generation was found to strongly contribute to I_{dark}. The observed decrease of E_{act} with increasing ΔV was interpreted as a non-negligible contribution of electric field effects. In order to separate these effects from quasi field independent mechanisms, a novel method was developed. The results obtained with this method confirmed the presence of a dominating diffusion current. Further, the contribution of dark currents caused by electric field effects was quantified. From the temperature dependence of Γ, it was concluded that the trap-assisted tunneling is the dominating electric-field dependent mechanism. However, a yet not quantified non-negligible contribution from band-to-band tunneling is anticipated. For a more detailed analysis of the contributing electric field effects, TCAD (Technology Computer Aided Design) simulations are required in order to evaluate the electric field distribution inside the GM-APD.

For the operation of the KETEK PM3350T STD at temperatures close to room temperature, the suppression of the charge carrier diffusion is necessary for a significant reduction of the dark count rate. In chapter 5, the developed approach to fulfill this requirement is discussed in detail.

Chapter 5

Suppression of the Diffusion Current

In chapter 4, the diffusion current $I_{diffusion}$ was identified to dominate the dark count rate of the KETEK PM3350T STD for temperatures $T \geq 10$ °C. In this chapter, an approach to suppress $I_{diffusion}$ of already fabricated SiPMs is presented. The main idea is to apply an external potential to the substrate of the SiPM and consequently to suppress the diffusion of minority charge carriers into the multiplication region. In section 5.1, the selected approach is motivated and the utilized experimental setup is discussed. Section 5.2 reviews the achieved suppression of the dark count rate and in section 5.3, the invariance of other SiPM parameters under the implemented device modification is confirmed.

5.1 Precedent considerations and experimental setup

The metrological methods used in chapter 4 do not provide any a priori indication of the type of charge carriers which are diffusing into the depletion region or any information on the region from which the carriers are diffusing into the active region.

In order to find a proper ansatz for the suppression of $I_{diffsuion}$, the following considerations were made: The electric field of a reverse biased GM-APD is oriented such that it counteracts a possible diffusion of majority holes from the p-layer or majority electrons from the n-layer into the active volume. Consequently, it was concluded that $I_{diffsuion}$ is caused by a diffusion of minority charge carriers.

One possible origin of the minority charge carrier diffusion are the quasi neutral regions at the lateral boundaries of the active region [95]. If this were the case, the probability of avalanche breakdowns along the edges of the active regions would be significantly increased. However, the results from the position resolved investigations of the dark count rate, which are presented in section 6.5, invalidate this assumption. Here, the highest DCR is observed in the central part of the micro-cells. Towards the edges, the dark count rate shows a significant decrease, which disagrees with a strong diffusion current from the peripheral regions.

Another possibility is the diffusion of minority charge carriers from the p- and n-layers. Since the doping concentration of the p-layer ($\approx 10^{19}$ cm^{-3}) is about two orders of magnitude higher than the doping concentration of the n-layer ($\approx 10^{17}$ cm^{-3}), it is assumed that the diffusion current of minority holes from the n-layer is much higher than the diffusion current of minority electrons from the p-layer.

The n-layer is implanted into a low-doped p-type substrate. For this reason an additional contribution of holes originating in the substrate and passing the n-layer towards the active region is expected.

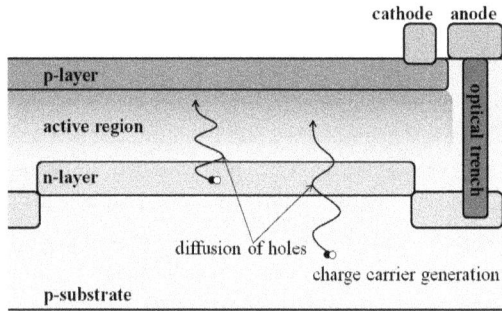

Figure 5.1: Sketch of the typical KETEK micro-cell structure. The diffusion of holes from the n-layer and the p-type substrate is indicated.

In figure 5.1, a sketch of a typical KETEK micro-cell structure is illustrated. The anticipated contributions from diffusion currents are indicated. The structure contains two pn-diodes. The active diode consists of the implanted p- and n-layers. This diode is operated in Geiger-Mode and generates a high-field active region for the detection of incoming photons. The backside diode consists of the implanted n-layer and the low doped p-type substrate. Since the substrate potential is not fixed, the operation conditions of this diode are not controlled. In order to counteract the charge carrier injection from the substrate to the active region, a fixed potential was applied to the backside of the KETEK PM3350T STD.

The additional substrate contact was realized by mounting the SiPM onto a golden pad of a PCB (printed circuit board) using conductive glue. The pad was then connected to an external voltage source in order to be able to change the electric potential at the substrate with respect to the n-layer. The utilized readout circuit is shown in figure 5.2.

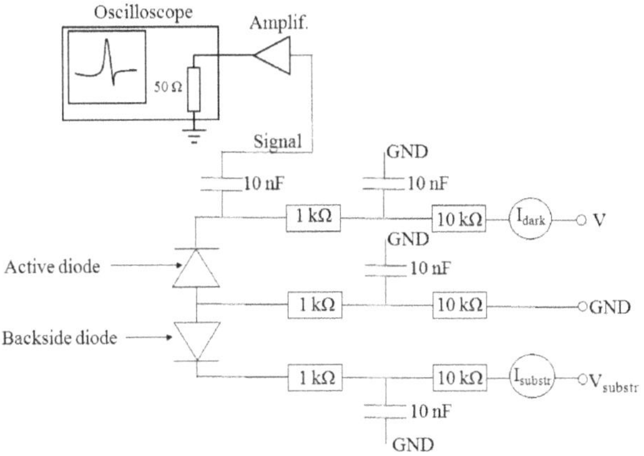

Figure 5.2: Circuit for the characterization of SiPM with an applied substrate potential.

The output of the SiPM is AC coupled to an oscilloscope to allow for measurements in pulsed mode. The currents through the active diode and the backside diode can be measured separately. With this setup, the backside diode can be operated at reverse biased conditions which reduces the concentration of free holes that may diffuse from the substrate towards the active region. The induced depletion region also reaches to some extend into the n-layer. For this reason, the diffusion of minority holes from the n-layer to the active region is also expected to be influenced by the substrate potential V_{substr}.

5.2 Impact of the substrate potential on the dark count rate

5.2.1 Investigations at room temperature

The following experiments were performed in a temperature controlled laboratory environment at $T = (21 \pm 1)$ °C. In figure 5.3(a), the dark count rate of the PM3350T STD at a fixed overvoltage of $\Delta V = 5$ V is shown as a function of the applied V_{substr}. The dark count rate decreases with decreasing V_{substr} and saturates for $V_{substr} \lesssim -1$ V.

In an additional experiment, the impact of V_{substr} on DCR was evaluated at different overvoltages. In figure 5.3(b), the dark count rate is plotted versus the overvoltage for $V_{substr} = 0$ V and $V_{substr} = -30$ V. The ratio of $DCR(V_{substr} = 0$ V$)$ and $DCR(V_{substr} = -30$ V$)$ is also displayed and is observed to decrease with increasing overvoltage. This observation is in agreement with the increasing contribution of high-field effects to the dark count rate with increasing ΔV, as shown in figure 4.15. This contribution is anticipated to not be affected by the applied substrate potential.

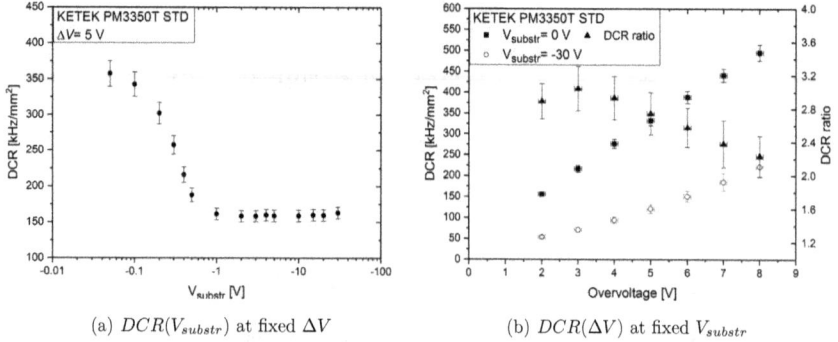

(a) $DCR(V_{substr})$ at fixed ΔV (b) $DCR(\Delta V)$ at fixed V_{substr}

Figure 5.3: Impact of the substrate potential V_{substr} on the dark count rate of the PM3350T STD.

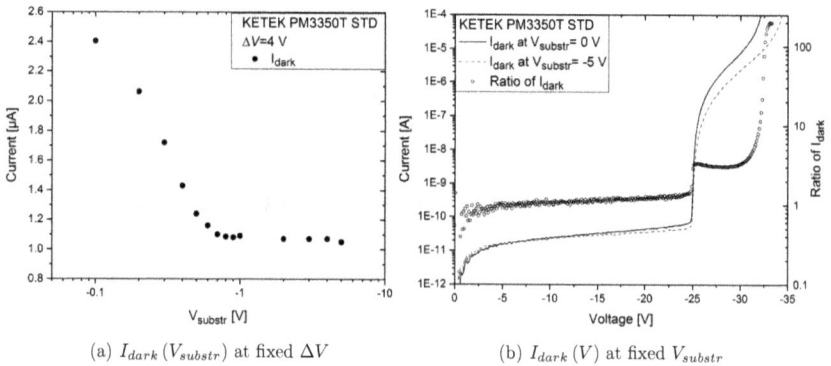

(a) $I_{dark}(V_{substr})$ at fixed ΔV (b) $I_{dark}(V)$ at fixed V_{substr}

Figure 5.4: Impact of the substrate potential V_{substr} on the dark current of the PM3350T STD.

Figure 5.4(a) shows the dependence of I_{dark} on V_{substr} for a fixed overvoltage. Analogous to the observed dependence of DCR on V_{substr}, I_{dark} decreases with decreasing V_{substr} and saturates for $V_{substr} \lesssim -1$ V. In figure 5.4(b), $I_{dark}(V)$ is shown at $V_{substr} = 0$ V and $V_{substr} = -5$ V. The achieved reduction of I_{dark} for $V > V_{BD}$ is evident in this figure. For operation voltages below the breakdown voltage, no significant impact of V_{substr} on I_{dark} was observed. In figure 5.5, the current I_{substr} which flows through the backside diode is shown as a function of the applied V_{substr}. As anticipated, the junction of the implanted n-layer and the p-type substrate shows a characteristic current-voltage behavior of a diode. However, the observed currents at reverse bias conditions exceed I_{dark} by roughly a factor of 10 at typical operation voltages of the SiPM. It is assumed that the dominating part of this "parasitic" current occurs due to high-ohmic connections between the substrate and the n-layer which are created at the edges of the SiPM during the wafer-dicing process. As will be discussed in section 5.3, I_{substr} does not show an impact on the photon detection capabilities or the noise parameters of the SiPM. Due to this result, I_{substr} was not investigated in detail within this work.

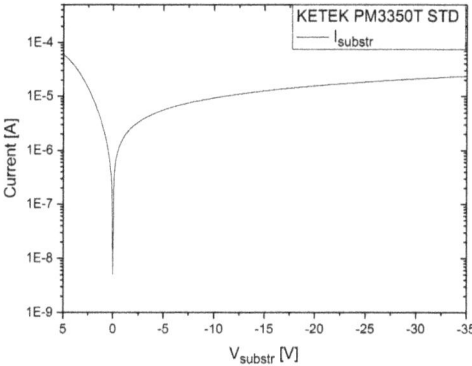

Figure 5.5: Parasitic current I_{substr} which flows during the operation of the backside diode at reverse bias conditions. This parasitic current does not affect the operation of the SiPM in pulsed mode.

Based on the observed behavior of the dark count rate with V_{substr}, the experimental setup for further investigations was simplified by short-circuiting the implanted p-layer and the p-type substrate of the KETEK PM3350T STD. This modification was done with regard to a simple implementation of the substrate potential on chip level. In the following, the SiPMs which were characterized with an applied substrate potential of $V_{substr} < -1$ V are termed by the suffix SP, such as the PM3350T STD-SP.

5.2.2 Investigations at different temperatures

As presented in the previous section, the dark count rate of the KETEK PM3350T STD was reduced for $T \approx 21$ °C by operating the backside diode of the SiPM structure at reverse bias conditions. In order to verify that the achieved improvement is indeed due to a suppressed diffusion current, the temperature dependence of the dark count rate at a fixed overvoltage of $\Delta V = 5$ V was measured for the KETEK PM3350T in STD and STD-SP mode. The results are shown in figure 5.6. For temperatures $T \leq -10$ °C, both modes show an equal DCR within the accuracy of the measurement. As discussed in section 4.3.3, the dominating contribution in this temperature range is the SRH-Generation enhanced by trap-assisted tunneling. A dominating contribution from $I_{diffusion}$ to the dark count rate is expected for temperatures higher than 10 °C. In this temperature range, an evidently lower DCR is observed for the PM3350T STD-SP. This result indicates, that the diffusion component is significantly suppressed by the applied substrate potential. The ratio of the dark count rates in both operation modes increases with increasing temperature, because $I_{diffusion}$ increases faster with T than $I_{gen+tat}$.

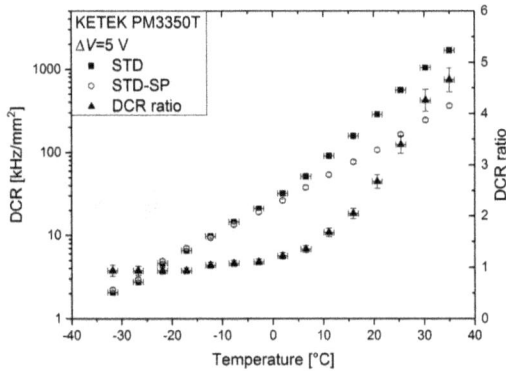

Figure 5.6: DCR of the PM3350T STD and of the PM3350T STD-SP versus temperature. For temperatures $T \geq 10$ °C, a significantly reduced DCR is evident for the PM3350T STD-SP.

Using the novel method presented in section 4.3.2, the temperature dependence of the dark currents of the PM3350T were analyzed in STD and in STD-SP mode. The measurements were performed in the temperature range from $T = 15$ °C to $T = 35$ °C, with an increment of 5 °C. The reconstructed dark currents are shown in figures 5.7(a) and 5.8(a). For both operation modes, the applied model is in agreement with the experimental data within an accuracy of < 10 % in the operation range of the SiPM.

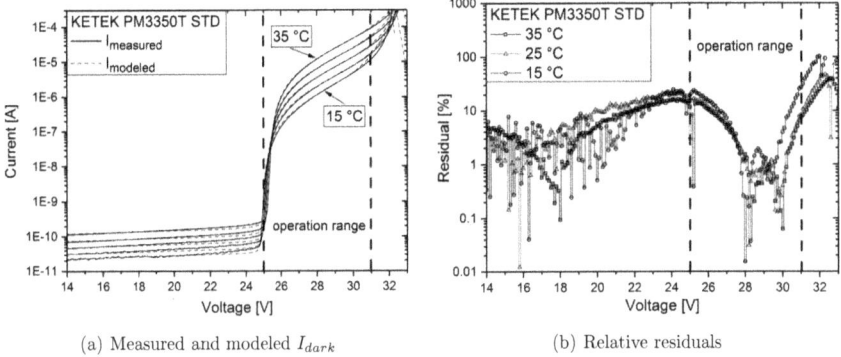

(a) Measured and modeled I_{dark} (b) Relative residuals

Figure 5.7: (a) Measured and modeled I_{dark} of the PM3350T STD at temperatures between 15 °C and 35 °C. (b) Relative residuals of the modeled I_{dark}.

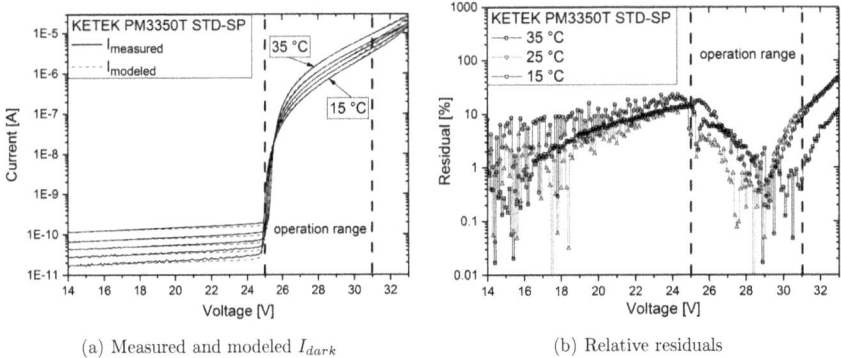

(a) Measured and modeled I_{dark} (b) Relative residuals

Figure 5.8: (a) Measured and modeled I_{dark} of the PM3350T STD-SP at temperatures between 15 °C and 35 °C. (b) Relative residuals of the modeled I_{dark}.

For operation voltages larger than 31 V, the investigated SiPM reaches the non-operation range and the used model cannot be applied anymore.

5. Suppression of the Diffusion Current

The dependence of the extracted I_{ini} on $(1/kT)$ is shown if figure 5.9 for the PM3350T STD and the PM3350T STD-SP, respectively. In order to extract E_{act}, the function 4.1 was fitted to the experimental data, with $\gamma = 3.2$ for the STD mode and $\gamma = 2.1$ for the STD-SP mode. For the PM3350T STD, an activation energy of $E_{act} = (1.12\pm0.03)$ eV was determined which reproduces the result obtained in section 4.3.3. For the PM3350T STD-SP, the extracted activation energy amounts to $E_{act} = (0.60 \pm 0.02)$ eV. This result confirms the strong suppression of the diffusion current by the substrate potential. However, the deviation of the extracted activation energy from the mid-bandgap energy of silicon indicates that the remaining contribution of the hole diffusion current is not entirely negligible. At this point it is assumed that the remaining contribution originates from non-depleted parts of the buried n-layer. In section 6.6, the contribution of the remaining $I_{diffusion}$ is discussed in more detail.

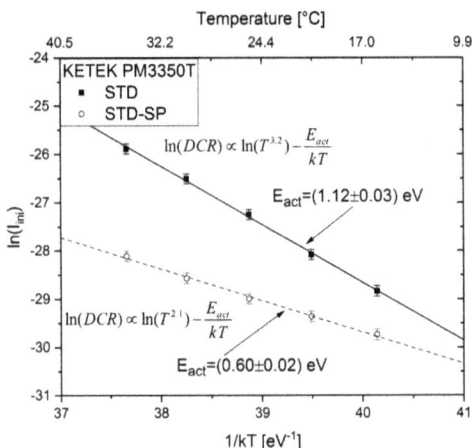

Figure 5.9: Arrhenius plot of I_{ini}. The activation energy of I_{ini} was reduced from 1.12 eV for the PM3350T STD to 0.60 eV for the PM3350T STD-SP.

Using the conventional method (see section 4.2), the activation energies of the dark count rate of the PM3350T STD-SP were determined in the temperature range from $T = -35$ °C to $T = +35$ °C. The corresponding Arrhenius plots are shown in figure 5.10(a) and the resulting activation energies are plotted in figure 5.10(b). For temperatures $T \leqslant -10$ °C, the extracted activation energy amounts to $E_{act} = (0.39 \pm 0.02)$ eV. This is in agreement with the results obtained for the PM3350T STD (see section 4.2).

For temperatures $T \geqslant 10$ °C, activation energies between (0.61 ± 0.02) eV for $\Delta V = 3$ V and (0.52 ± 0.02) eV for $\Delta V = 6$ V were determined. This is in agreement with the results shown in figure 5.9.

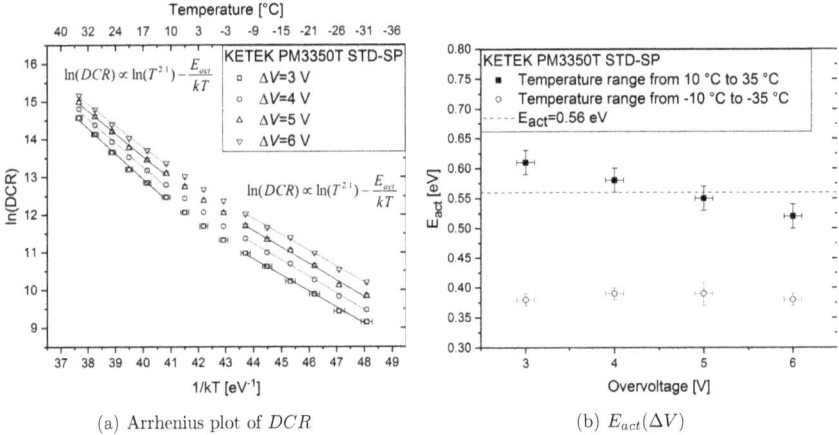

(a) Arrhenius plot of DCR

(b) $E_{act}(\Delta V)$

Figure 5.10: (a) Arrhenius plots of the dark count rate of the PM3350T STD-SP at different overvoltages. (b) Resulting activation energies versus the overvoltages.

5.3 Impact of the substrate potential on other SiPM parameters

In the previous section, the impact of V_{substr} on the dark count rate was discussed. In this section, the impact of the substrate potential on other SiPM parameters is evaluated.

The pulse shape of the SiPMs does not change with the application of the substrate potential. This is demonstrated in figure 5.11(a) at $\Delta V = 5$ V. Figure 5.11(b) shows the corresponding pulse-height spectra, which were acquired at dark conditions. Also here, no difference is observed between the two operation modes.

The invariance of the relative gain under the application of V_{substr} is displayed in figures 5.12(a) and 5.12(b). Accordingly, the breakdown voltage is also not influenced by the substrate potential. The absolute gain is also not influenced by V_{substr}. In figure 5.13(a), the pulse shapes of the saturated (all micro-cells fire simultaneously) PM3350T in the STD mode and the STD-SP mode are shown at $\Delta V = 5$ V. In figure 5.13(b), the invariance of the gain G under the applied substrate potential is shown.

Figures 5.14(a) and 5.14(b) confirm that the crosstalk probability of the PM3350T STD and the PM3350T STD-SP is indistinguishable within the accuracy of the measurement.

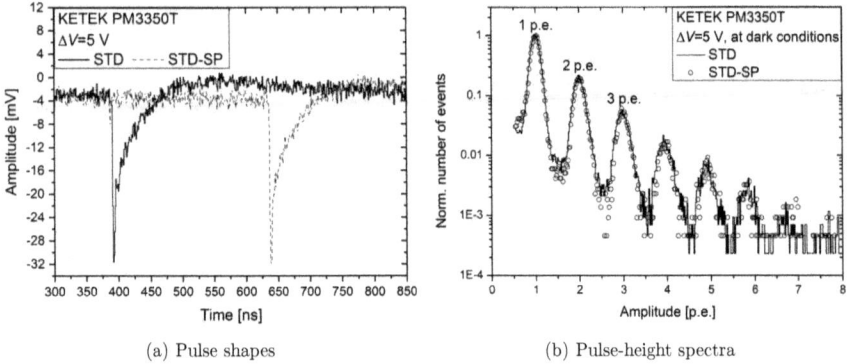

(a) Pulse shapes

(b) Pulse-height spectra

Figure 5.11: Impact of the substrate potential on the SiPM pulse shape.

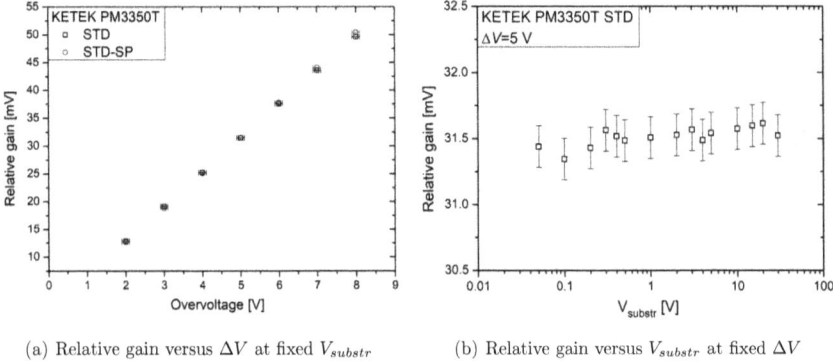

(a) Relative gain versus ΔV at fixed V_{substr}

(b) Relative gain versus V_{substr} at fixed ΔV

Figure 5.12: Impact of substrate potential on the relative gain.

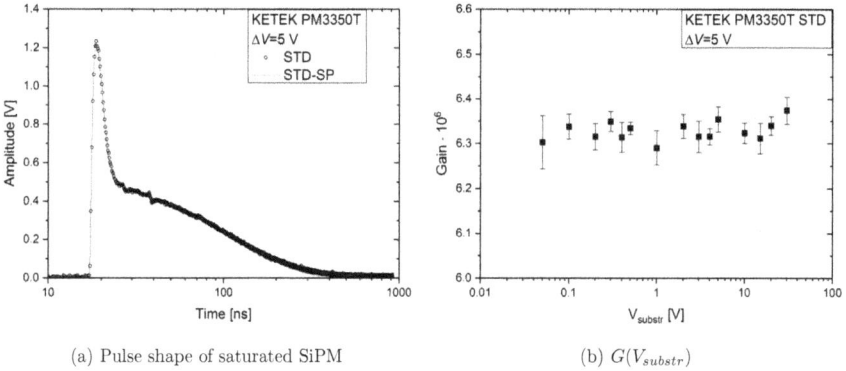

(a) Pulse shape of saturated SiPM

(b) $G(V_{substr})$

Figure 5.13: Impact of substrate potential on the absolute gain G.

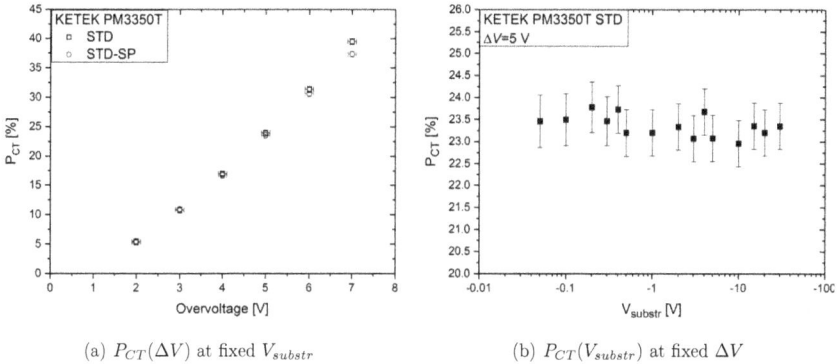

(a) $P_{CT}(\Delta V)$ at fixed V_{substr}

(b) $P_{CT}(V_{substr})$ at fixed ΔV

Figure 5.14: Impact of substrate potential on the optical crosstalk probability P_{CT}.

In figure 5.15(a), the CCDFs (see section 3.7) of pulses subsequent to a primary dark pulse are shown for the PM3350T STD and the PM3350T STD-SP. The suppressed DCR is reproduced in a curvature change of P_{tot}^*. In figure 5.15(b), the probability of correlated pulses is plotted versus the overvoltage. For overvoltages < 5 V, no variation of P_{CP} is observed. For overvoltages ≥ 5 V, a slight increase of P_{CP} for the PM3350T STD-SP is evident. It is assumed that the observed variation is caused by the following:

Due to a decreased dark count rate of the PM3350T STD-SP, the delay-times at which the dark count rate dominates the CCDF are shifted to higher values. Consequently, the fitting range for the exponential fit reduces, because the length of the signal trace is limited to 1 μs. This has an impact on the quality of the fit and introduces an uncertainty to the extracted P_{CP}.

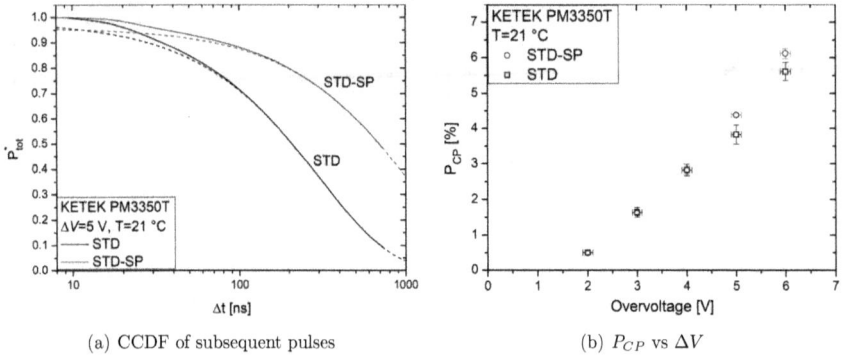

(a) CCDF of subsequent pulses

(b) P_{CP} vs ΔV

Figure 5.15: Impact of substrate potential on the probability of correlaed pulses P_{CP}.

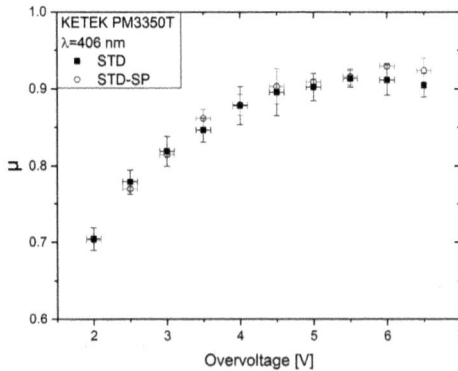

Figure 5.16: Impact of the substrate potential on the photon detection efficiency.

One of the most crucial SiPM parameters is the photon detection efficiency (PDE). For this reason, the invariance of PDE under the application of V_{substr} is mandatory for the implementation of the substrate potential in batch production. Since the geometrical efficiency of the PM3350T STD and STD-SP is the same, the comparison of the average number of detected photons μ at a fixed output of the light source is sufficient for the evaluation of PDE. In figure 5.16, the comparison of $\mu(\Delta V)$ at 406 nm is shown. No impact of V_{substr} on $\mu(\Delta V)$ and hence no impact on PDE is observed within the uncertainties.

5.4 Conclusion

In this chapter, an innovative approach to suppress the identified diffusion current was presented. This approach is based on the application of an electric potential to the substrate of the SiPM and hence on the operation of the backside diode at reverse bias conditions. With this method, majority charge carriers from the p-type substrate are prevented from diffusing through the n-layer and into the active region of micro-cells. This led to a significant reduction of the dark count rate of the KETEK PM3350T STD for temperatures larger than 10 °C.

The temperature dependence of the dark count rate of the PM3350T STD-SP indicates that the diffusion current is not suppressed entirely by the applied external potential. It is assumed that the remaining $I_{diffusion}$ originates from the non-depleted regions of the buried n-layer. In section 6.6, this hypothesis is discussed in more detail.

Chapter 6

Spatially Resolved Dark Count Rate

The experimental methods discussed in the previous chapters have one common disadvantage: No spatial information on the generation of dark pulses is accessible, since the SiPM output signal is the sum of the output signals of each firing micro-cell. In this chapter, a method is presented which uses the effect of hot carrier luminescence (see section 2.6) to achieve a 2D spatially resolved determination of the SiPM dark count rate. In section 6.1, the used experimental setup is described. In section 6.2, the existence of micro-cell regions with different dark count rates is introduced, and the investigation of the regions with an enhanced DCR (hotspots) is motivated. In sections 6.3 and 6.4, the developed experimental method for the hotspot analysis is described. The impact of the substrate potential on the dark count rate in different regions is investigated in section 6.5. In section 6.6, the mechanisms which are responsible for the occurrence of hotspots are identified by a spatially resolved determination of the activation energy.

6.1 Experimental setup

For the measurements, the SiPM under test was placed inside a dark box. The light emitted during the avalanche breakdowns of micro-cells was detected with the low-light-level CCD camera *Andor, Clara*, which was attached to the optical microscope *Mitutoyo FS70 S/N*. The camera was used in extended mode for the near-infrared (NIR) range. The quantum-efficiency curve of the CCD is shown in figure 6.2 [96]. In figure 6.1, a sketch of the used setup is shown. All data was background corrected by subtracting a CCD image with the SiPM bias voltage set to zero. The measurements were performed in a temperature stabilized environment at (21 ± 1) °C. For investigations at different temperatures, the SiPMs were mounted onto a TEC as shown in figure 4.2.

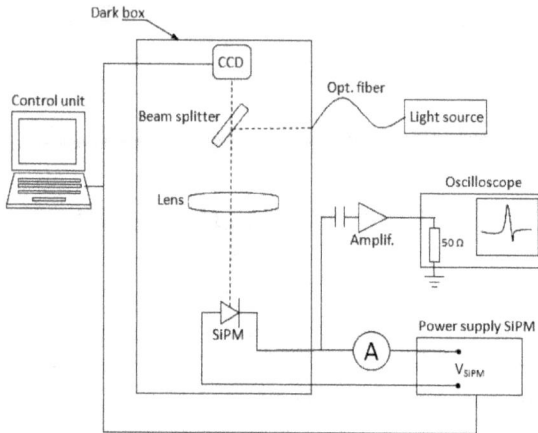

Figure 6.1: Sketch of the experimental setup for the detection of hot carrier luminescence.

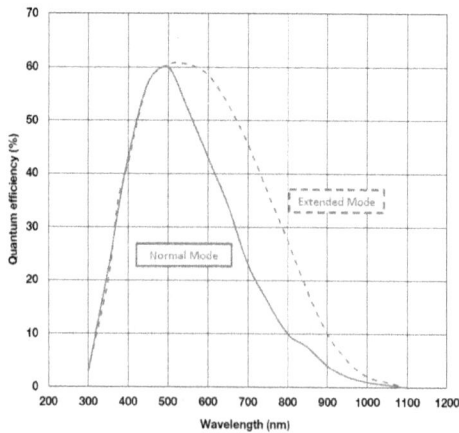

Figure 6.2: Quantum efficiency curve of the Clara CCD camera. Taken from [96].

6.2 Existence of hotspots

In 2009, Frach et al. [97] reported on a digital SiPM (dSiPM), which was fabricated by PHILIPS in CMOS technology. The dSiPM provides the possibility of a separated readout of individual micro-cells. Using this feature, the group was able to generate a map which contained the DCR values of each micro-cell of the dSiPM. It was concluded that by excluding 5 % of the most active micro-cells, the dark count rate can be reduced by up to an order of magnitude. However, no detailed information on the underlying physics of micro-cells with a higher DCR was reported.

In a conventional (analog) SiPM, the micro-cells are connected to a common readout. For this reason, it is not possible to measure the dark count rate of each micro-cell individually by using the conventional methods described in section 3.6. In this work, a novel method was developed which provides access to a spatially resolved characterization of the dark count rate of SiPMs. This method is based on the effect of hot carrier luminescence, which is discussed in section 2.6.1. Light, which is emitted by the SiPM micro-cells during avalanche breakdowns is detected with a low-light-level CCD. Correlating the emitted light intensity and the dark count rate, a DCR map with a sub-micro-cell resolution is generated.

Figure 6.3(a) shows a photo of a section of the PM3350T STD at zero bias voltage under ambient light conditions. In figure 6.3(b), the light emission image of the same section is shown without external light at $\Delta V = 5.4$ V. Sub-micro-cell regions with a significantly higher light intensity (hotspots) are clearly evident within the active areas of the SiPM micro-cells. Figure 6.4 shows to what fraction each micro-cell contributes to the overall emitted light intensity. For this figure, the CCD pixel amplitudes of the light emission image 6.3(b) were integrated for each micro-cell and normalized with the total light intensity from 221 micro-cells. Micro-cells containing a hotspot, contribute with up to 7 %, whereas an average contribution of approximately 0.45% is expected. In section 6.4, the correlation of the light emission intensity and the dark count rate is discussed. Considering this correlation, the obtained result indicates that a fraction of micro-cells exceeds the expected DCR by more than an order of magnitude. This is in agreement with the observations reported in [97]. The enhanced contribution of certain micro-cells is attributed to the existence of hotspots within the active area of the SiPM. The goal of this work is to generate a deeper understanding of the origin of hotspots in order to suppress their contribution to the dark count rate of SiPMs.

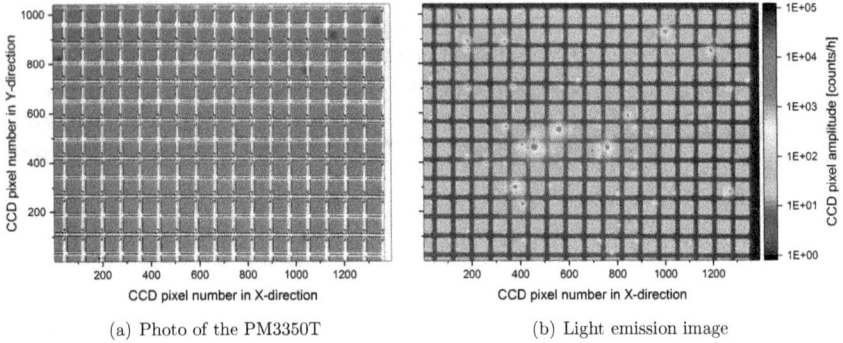

(a) Photo of the PM3350T

(b) Light emission image

Figure 6.3: (a) CCD image of the PM3350T STD at ambient light conditions. (b) Light emission image at dark ambient conditions with $\Delta V = 5.4$ V and $t_{exp} = 4$ h.

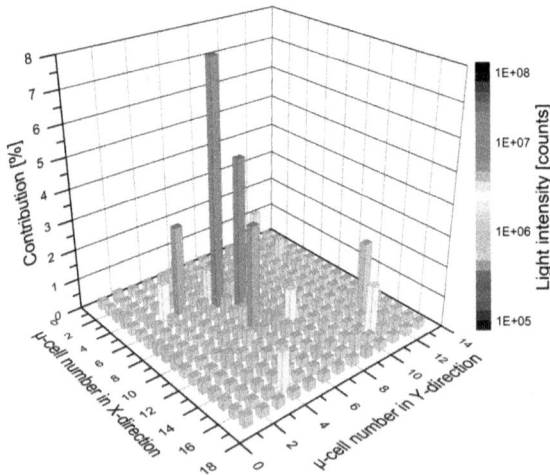

Figure 6.4: Contribution of each micro-cell from figure 6.3(b) to the overall emitted light intensity.

6.3 Hotspot analysis - Algorithm

In this section, the developed metrological method for the analysis of hotspots is described by taking the example of the PM3350T STD-SP. The number of hotspots and the light intensity emitted from each hotspot is determined. The emitted light intensity is correlated with the dark count rate and the contribution of hotspots to the dark count rate is evaluated.

6.3.1 Determination of $I_{glowing}$ and $I_{hotspots}$

The camera signal is divided into two main contributions. The first contribution is attributed to the homogeneous emission of light from every micro-cell without hotspots (bluish regions in figure 6.3(b)). This contribution is called "glowing intensity" $I_{glowing}$ and is defined as the sum of all CCD pixel amplitudes below the hotspot threshold $T_{hotspots}$. The second contribution is due to hotspots and is defined as the sum of all pixel amplitudes above $T_{hotspots}$. This quantity is called "hotspots intensity" $I_{hotspots}$. Equations 6.1 and 6.2 give a formal definition of both quantities, where $A_{i,j}$ is the experimentally determined CCD pixel amplitude at position (i,j).

$$I_{glowing} = \sum_{i=1}^{n} \sum_{j=1}^{m} A_{i,j} \text{ , for } A_{i,j} \leq T_{hotspots} \tag{6.1}$$

$$I_{hotspots} = \sum_{i=1}^{n} \sum_{j=1}^{m} A_{i,j} \text{ , for } A_{i,j} > T_{hotspots} \tag{6.2}$$

Figure 6.5 shows an example of a CCD pixel amplitude spectrum of the PM3350T STD-SP. The spectrum is continuous and does not provide a feature that allows to define a general $T_{hotspots}$. For this reason, an arbitrary method is introduced to distinguish between $I_{glowing}$ and $I_{hotspots}$. In the presented method, $T_{hotspots}$ is determined individually for each SiPM and operation voltage. The left part of the spectrum is attributed to the light emission from hotspot-free regions. This part is fitted with a Gaussian and $T_{hotspots}$ is set at 4 standard deviations from the mean of the distribution, indicated by the vertical line in figure 6.5. During the data acquisition, cosmic rays interact with the CCD and generate additional noise signals with CCD pixel amplitudes comparable to hotspots. In figure 6.5, the influence of cosmic rays on the CCD signal is evident as the non-Gaussian tail of the distribution at zero bias voltage (dashed curve). The contribution of the total camera noise I_{noise} to $I_{glowing}$ is shown in figure 6.6. This contribution decreases with the increasing overvoltage and is below 5 % at $\Delta V = 5.4$ V.

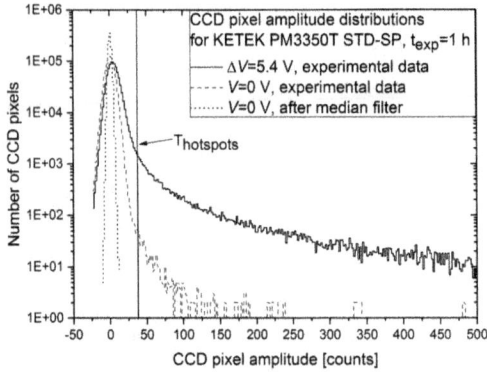

Figure 6.5: Examples of CCD pixel amplitude distributions.

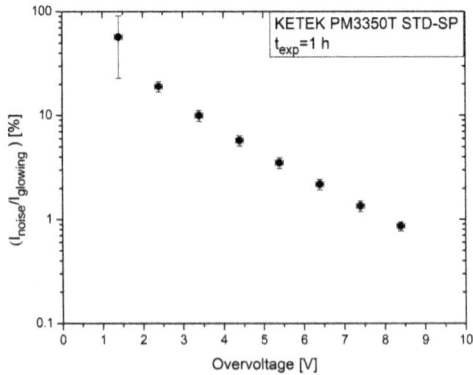

Figure 6.6: Contribution of the CCD noise I_{noise} to $I_{glowing}$ for $t_{exp} = 1$ h.

Figure 6.7 shows an example of the pixel amplitudes from one CCD row, which contains a hotspot and a cosmic ray peak. It is evident from this figure that the chosen threshold $T_{hotspots}$ provides a clear separation between hotspots and hotspot-free regions. Considering the fact that cosmic ray signals appear as peaks, which are much narrower (about 1 to 3 CCD pixels) than hotspots (> 10 CCD pixels), a 1D median filter was used to exclude these peaks from the analysis. The filter successively compares the amplitudes of a CCD pixel and its 6 nearest right and 6 nearest left neighbors. The experimentally determined $A_{i,j}$ is then changed to the median amplitude of the analyzed pixels. In this way, sharp peaks are suppressed. The dotted curve in figure 6.5 shows the CCD pixel amplitude distribution at zero bias voltage after the application of the median filter. The suppression of the cosmic ray peaks is clearly evident.

The median filter significantly modifies the amplitude and shape of hotspots, as indicated by the solid line in figure 6.7. In order to determine the hotspot contribution to the total light intensity, a reconstruction of the hotspot peaks is necessary. For this purpose, the distribution of the filtered CCD pixel amplitudes $A_{i,j}^{filt}$ is analyzed and the threshold T_{filt} is set analogous to the method described for $T_{hotspots}$. The reconstructed CCD pixel amplitudes $A_{i,j}^{rec}$ at the positions of hotspots are determined according to equation 6.3. The dashed line in figure 6.7 illustrates an example of a reconstructed hotspot.

$$A_{i,j}^{rec} = \begin{cases} A_{i,j} & \text{for } A_{i,j} > T_{hotspots} \text{ and } A_{i,j}^{filt} > T_{filt} \\ 0 & \text{for } A_{i,j} \leq T_{hotspots} \text{ or } A_{i,j}^{filt} \leq T_{filt} \end{cases} \tag{6.3}$$

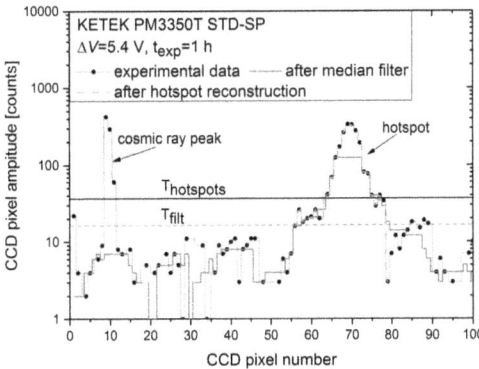

Figure 6.7: Example of a hotspot and a cosmic ray peak in 1D.

99

6.3.2 Counting of hotspots

The number of hotspots and the light emission intensity of each hotspot is determined in three steps. In the first step, the positions of hotspot centers is determined by identifying the coordinates of local amplitude maxima of the CCD pixels. In the second step, the lateral expansion of hotspots along the i-coordinate is determined as the distance between the hotspot center and the fist CCD pixel with $A_{i,j}^{rec} = 0$. Since the spatial intensity distribution of individual hotspots can be described by a 2D Gaussian, the expansion along the i-coordinate is sufficient. The CCD pixel amplitudes $A_{i,j}^{rec}$ attributed to one hotspot are summed up and interpreted as the light emission intensity of the hotspot. This quantity is called $I_{hotspots}^{single}$. In the last step, the detected hotspots are sorted in a decreasing order of $I_{hotspots}^{single}$. The quantity $N_{hotspots}$ is defined by 90 % of the brightest hotspots. The additional cut is implemented in order to suppress false positives with small CCD pixel amplitudes.

The presented method is applicable under the condition of well separated hotspots. The emitted light intensity increases with ΔV as a consequence of the increasing DCR and gain of the SiPM (see equation 6.4). The FWHM (full width at half maximum) of the lateral hotspot expansion does not show a significant increase with ΔV [98]. However, an absolute increase of the lateral expansion of hotspots is observed with the overvoltage, as displayed in figure 6.8.

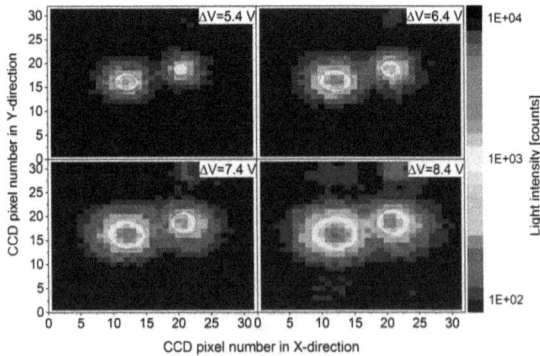

Figure 6.8: Example of the lateral expansion of hotspots with the overvoltage for the PM3350T STD-SP ($t_{exp} = 1$ h). One CCD pixel corresponds to 3.23 μm.

For the case that $A_{i,j}^{rec}$ does not drop below zero in between the local maxima, the hotspots cannot be separated. But, because the observed number of hotspots is much lower than the number of SiPM micro-cells, the uncertainty caused by this effect is neglected.

The number of detected hotspots depends on the choice of $T_{hotspots}$. In figure 6.9, $N_{hotspots}$ is shown as a function of $T_{hotspots}$ for PM3350T STD-SP at $\Delta V = 5.4$ V. The vertical line indicates $T_{hotspots}$ that was chosen in this work.

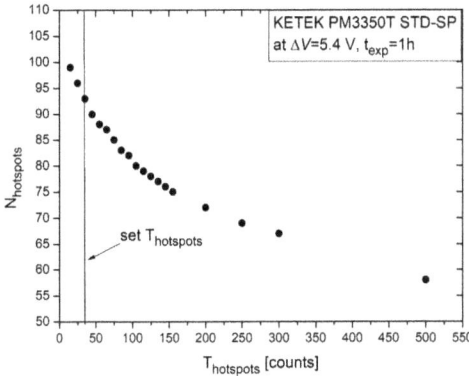

Figure 6.9: $N_{hotspots}$ versus $T_{hotspots}$. The vertical line indicates $T_{hotspots}$ according to the method described in section 6.3.1.

In figure 6.10, $N_{hotspots}$ is shown as a function of overvoltage for the PM3350T STD-SP. The increase of $N_{hotspots}$ is attributed to the increase of the avalanche triggering probability P_{trigg}. At low overvoltages only crystal defects at the position of the maximum electric field are expected to contribute significantly to the dark count rate. With increasing overvoltage, the electric field and hence P_{trigg} increases. The contribution of crystal defects located in regions of an initially low electric field is consequently enhanced. $N_{hotspots}$ is expected to reach saturation at overvoltages at which the position dependent P_{trigg} is saturated. In figure 6.11, the relative increase of $N_{hotspots}$ and of $PDE(\lambda = 406$ nm$)$ is shown as function of ΔV. The saturation behavior of both quantities is in agreement within the uncertainties. The absorption coefficient in silicon at $\lambda = 406$ nm amounts to $\alpha \approx 7.55 \cdot 10^4$ cm^{-1}. Consequently, figure 6.11 indicates that the detected hotspots are located within a depth of approximately 0.5 μm.

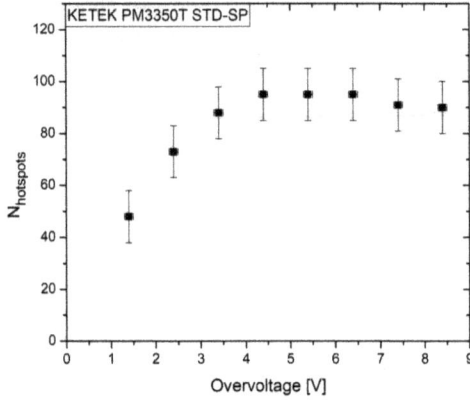

Figure 6.10: $N_{hotspots}$ versus the overvoltage for the PM3350T STD-SP ($t_{exp} = 1$ h).

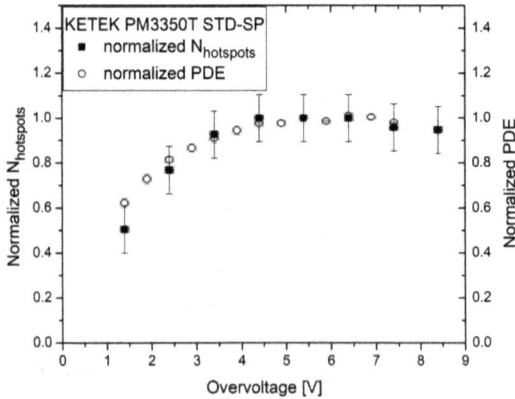

Figure 6.11: Comparison of the saturation behavior of $N_{hotspots}$ and $PDE(\lambda = 406$ nm) with overvoltage.

102

6.4 Correlation of the light emission intensity and the dark count rate

Equation 6.4 gives a general expression of the detected fraction I of the total emitted light intensity. Here, t_{exp} is the exposure time, G is the gain of the SiPM, DCR is the overall dark count rate, η is the correction factor accounting for the effect of optical crosstalk (see equations 4.4 and 4.3), $PDE_{CCD}(\lambda)$ is the detection efficiency of the CCD camera and $\kappa(\lambda)$ is the number of emitted photons per electron-hole pair.

$$I = t_{exp} \cdot G \cdot DCR \cdot \eta \cdot \int_\lambda PDE_{CCD}(\lambda) \cdot \kappa(\lambda) \, \mathrm{d}\lambda \qquad (6.4)$$

The method which was used for the determination of DCR does not take into account the amplitudes of the detected pulses, as described in section 3.6. By the effect of optical crosstalk, secondary photons emitted during a primary dark pulse initiate additional avalanches in neighboring micro-cells. The additionally firing micro-cells then again produce secondary photons, which are detected by the CCD. For a correlation of I and DCR, the factor η is necessary to account for the number of "simultaneous" crosstalk pulses following an initial dark pulse (see section 4.2)

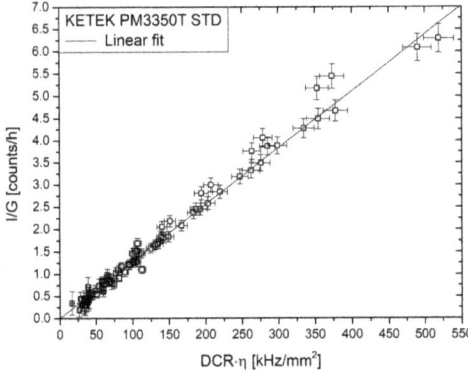

Figure 6.12: Linear relation of I/G and the $DCR \cdot \eta$.

In figure 6.12, the ratio I/G is shown as a function of $DCR \cdot \eta$. The data points were obtained by a variation of ΔV. The observed linear correlation between the emitted light intensity and the dark count rate is in agreement with the observations reported in [63] for an avalanche photo diode (see figure 2.18). This result establishes the possibility to generate a spatially resolved DCR map of the SiPM.

The contribution R of hotspots to the overall DCR is determined using equation 6.5. Analogous to $N_{hotspots}$, R depends on the choice of the hotspot threshold $T_{hotspots}$. Figure 6.13 demonstrates the impact of $T_{hotspots}$ on R for the PM3350T STD-SP at $\Delta V = 5.4$ V

$$DCR_{hotspots} = \frac{I_{hotspots}}{I_{hotspots} + I_{glowing}} \cdot DCR_{tot} = R \cdot DCR_{tot} \quad (6.5)$$

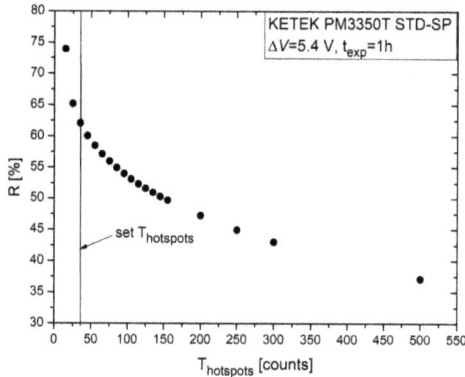

Figure 6.13: Contribution R of hotspots to the total dark count rate versus $T_{hotspots}$.

The presented method is only applicable within the operation range of the SiPM, which extends to $\Delta V \approx 7.5$ V for the PM3350T STD-SP. The overvoltage range from $\Delta V \approx 7.5$ V to $\Delta V \approx 10.5$ V is called "transition range". Here, contributions like high field effects and enhanced probabilities of crosstalk pulses and of correlated pulses start to dominate the performance of the SiPM. The mentioned effects contribute mostly to $I_{glowing}$ which leads to a decrease of R. Additionally, hotspots with a high light emission intensity start to overlap, which is enhanced by the increased P_{CT} and P_{CP}. For $\Delta V \gtrsim 10.5$ V, a rapid increase of the dark current is observed. In this "non-operation range", the characterization of the SiPM is not possible. In figure 6.14, an example of $I_{dark}(\Delta V)$ is shown for the PM3350T STD. The discussed overvoltage ranges are marked. The overvoltages and the width of the described ranges depend on the SiPM design and might show significant variations even within the same batch.

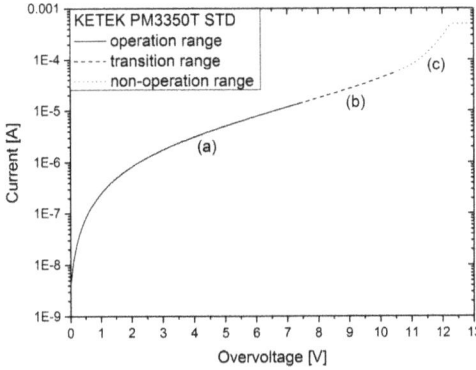

Figure 6.14: $I_{dark}(\Delta V)$ of the PM3350T STD divided into three overvoltage ranges: (a) operation range, (b) transition range and (c) non-operation range.

As shown in figure 6.15, the hotspot contribution to the overall DCR for the investigated PM3350T STD-SP amounts to $R \approx 55\ \%$. Based on this result, a DCR reduction by a factor larger than 2 is expected for the case of a complete hotspot suppression. Furthermore, R is observed to be independent of the overvoltage. In order to understand this observation, the position dependent $DCR(r, F, \Delta V)$ is approximated as a sum of contributions from n participating mechanisms (see equation 6.6). Here, r is the position and F is the electric field strength. The contribution of the i-th mechanism is described as a product of the respective free charge carrier generation rate $f_i(r, F)$ and the respective avalanche initiation probability $P_i^{trigg}(r, \Delta V)$.

$$DCR = \sum_{i=1}^{n} f_i(r, F) \cdot P_i^{trigg}(r, \Delta V) \tag{6.6}$$

Considering this equation, the independence of R on ΔV within the operation range can be explained by the following hypothesis: "For the PM3350T STD-SP, the generation of dark pulses at the positions of hotspots and in hotspot-free areas is dominated by the same mechanism at $T \approx 21\ °C$. Otherwise, the difference between the contributing f_i at different positions would lead to a dependence of R on the electric field conditions." In section 6.6 this topic is reviewed in more detail.

105

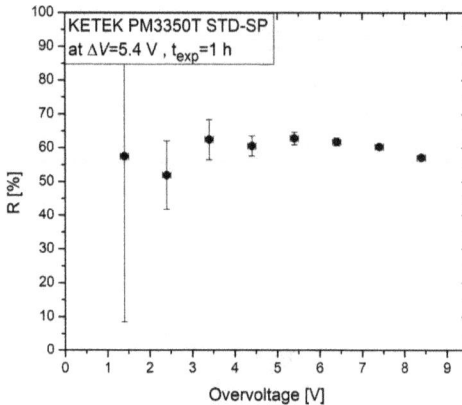

Figure 6.15: The contribution R of hotspots to the total dark count rate versus the overvoltage for the PM3350T STD-SP.

6.5 Impact of the substrate potential

In a first optimization step of the KETEK PM3350T STD, the contribution from minority charge carrier diffusion to the dark count rate was suppressed by the application of a substrate potential (see chapter 5). In this section, the impact of the substrate potential on the regions with and without hotspots is reviewed. For this purpose, the light emission intensities and distributions in the regions of interest are compared for the PM3350T STD and the PM3350T STD-SP.

In figures 6.16(a) and 6.16(b), the light emission images of the PM3350T in STD and STD-SP mode are shown at $\Delta V = 4$ V. In both cases, regions with a homogeneous light emission can be observed. This contribution is attributed to the parameter $I_{glowing}$ as described in section 6.3.

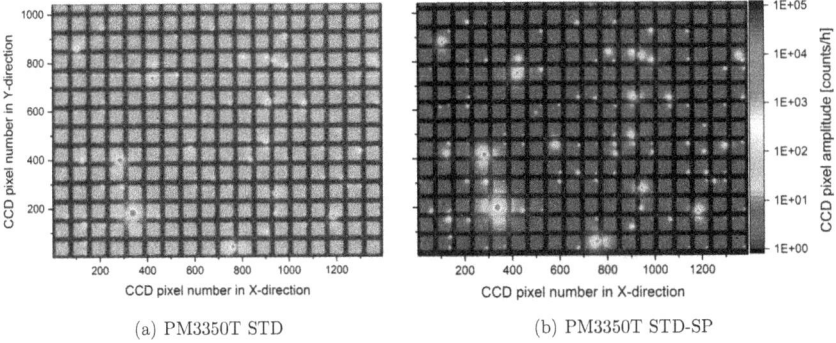

(a) PM3350T STD

(b) PM3350T STD-SP

Figure 6.16: Light emission images of the PM3350T in (a) STD mode and (b) in STD-SP mode at $\Delta V = 4$ V, $t_{exp} = 1$ h and $T = 20$ °C.

The corresponding contour plots of the average micro-cell are shown in figure 6.17. Here, only micro-cells without hotspots were considered. The contained sharp, high intensity peaks are attributed to cosmic particles hitting the CCD during the acquisition. In this analysis, these peaks were not excluded in order to prevent any modification of the light distribution. In figure 6.18, the corresponding CCD pixel amplitudes along the center of the contour plots are shown in 1D.

The assumption is made that for micro-cells without hotspots, the free charge carrier generation rate (see equation 6.6) is distributed uniformly over the active area. Consequently, the light intensity displays the spatial distribution of P_{trigg} and hence the distribution of PDE. In the central part of the micro-cell, the distribution of the light intensity has a plateau. In this region, the electric field is independent of the lateral directions at a fixed depth within the micro-cell. Towards the edges of the micro-cell, the light intensity and hence the photon detection efficiency show a rapid decrease. This effect is connected with the degradation of the electric field strength in lateral directions for the purpose of preventing an enhanced rate of avalanche breakdowns at the micro-cell edges.

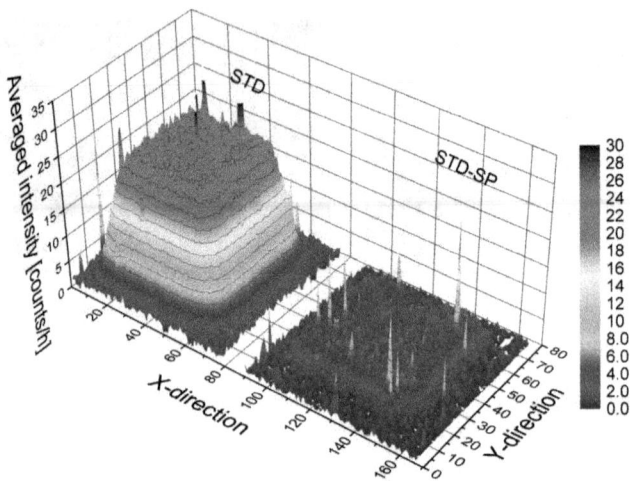

Figure 6.17: Contour plots of the average light emission from one micro-cell of the PM3350T STD and the PM3350T STD-SP ($t_{exp} = 1$ h, $\Delta V = 4$ V, $T = 20$ °C).

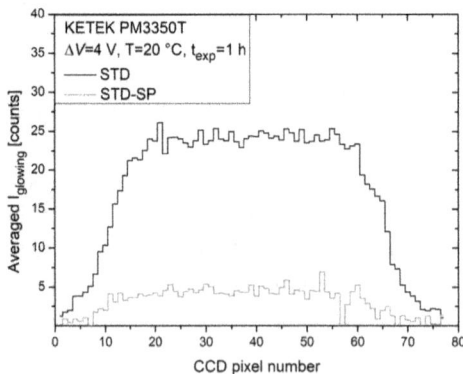

Figure 6.18: Average light emission from one micro-cell of the PM3350T STD and STD-SP in 1D. The data points correspond to the cross-section along the center of the contour plots in figure 6.17.

The obtained results lead to the conclusion that the suppression of the diffusion component reduced the light emission and hence the dark count rate of micro-cells without hotspots by a factor of 5. The homogeneous reduction of the light distribution within the micro-cells supports the assumption that the diffusion of minority charge carriers into the multiplication region occurs from the backside of the structure and is spread homogeneously over the active area of the micro-cells. Enhanced contributions from low-doped regions at the edges of the active area are excluded.

In order to investigate the impact of V_{substr} on hotspots, the method described in section 6.3 was applied to the light emission images 6.16(a) and 6.16(b). In figure 6.19(a), the coordinates of the detected hotsptots are shown. No impact of V_{substr} on the number and lateral position of hotspots is observed. The slight shift of the hotspot coordinates along the Y-direction is attributed to the fact that the SiPM under test was moved between the exposures in order to switch from STD to STD-SP mode. In figure 6.19(b), $I_{hotspots}^{single}$ of the detected hotspots is shown for the PM3350T in STD and STD-SP mode, respectively. From this plot, it is concluded that the applied V_{substr} does not have any impact on the occurrence of hotpot and their contribution to the dark count rate. Consequently, the process responsible for the occurrence of hotspot is not related with charge carrier diffusion. In section 6.6, the mechanism identified to be responsible for hotspots is reviewed in detail.

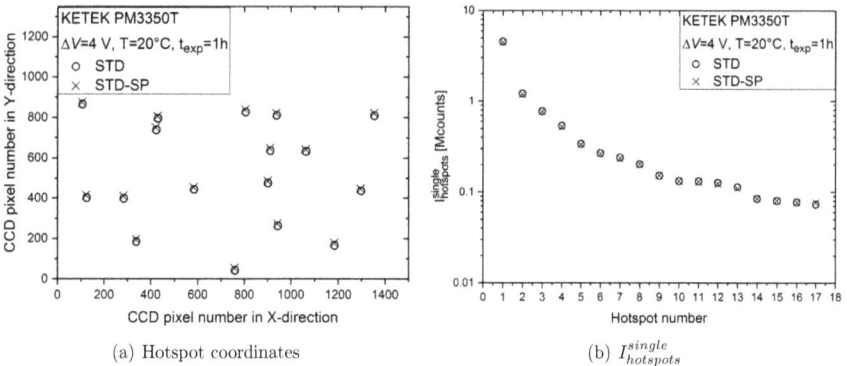

(a) Hotspot coordinates

(b) $I_{hotspots}^{single}$

Figure 6.19: Impact of the substrate potential on (a) the coordinates of hotspots and on (b) the light emission intensity of hotspots.

6.6 Spatially resolved activation energy

The experimental method presented in the previous sections provides access to the dark count rate in every part of the active region. In this section, this method is extended by the determination of spatially resolved activation energies. The idea is to use the temperature dependence of the emitted light intensity, analogous to the conventional method discussed in section 4.2. One objective of this section, is to evaluate the level, to which the diffusion of minority charge carriers is suppressed by V_{substr}. Another objective is to reveal the physical mechanism which is responsible for hotspots.

As mentioned in section 6.5, the dark count rate is a function of the avalanche triggering probability and the charge carrier generation rate of the contributing physical mechanisms. In order to probe these mechanisms, a fixed P_{trigg} is mandatory. As shown in section 3.5, no dependence of $PDE(\Delta V)$ on temperature is expected. For this reason, no corrections have to be applied to the detected light emission for different temperatures as long as the analysis is performed at a fixed overvoltage.

In figure 6.20(a), the CCD pixel amplitude map of 4 micro-cells of a PM3350T STD is shown. The data was acquired at $\Delta V = 4$ V, $T = 30$ °C and $t_{exp} = 1$ h. The active regions are clearly separated from the periphery and two of the micro-cells contain hotspots of different intensities.

The transition from the amplitude map to an activation energy map is realized by a pixel-wise analysis of the CCD pixel amplitudes at two distinct temperatures. Analogous to the dark count rate, the detected light emission intensity I is approximated as shown in equation 6.7. The activation energy of each CCD pixel is determined as depicted in equation 6.8. Here, $I_{1,2}$ describes the light intensity at $T_{1,2}$.

$$I = I_0 \cdot \exp\left(-\frac{E_{act}}{kT}\right) \tag{6.7}$$

$$E_{act} = -\ln\left(\frac{I_1}{I_2}\right) \cdot \left(\frac{1}{kT_1} - \frac{1}{kT_2}\right)^{-1} \tag{6.8}$$

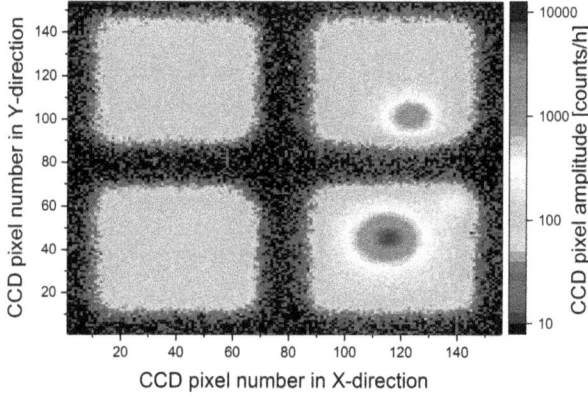

(a) Light emission image

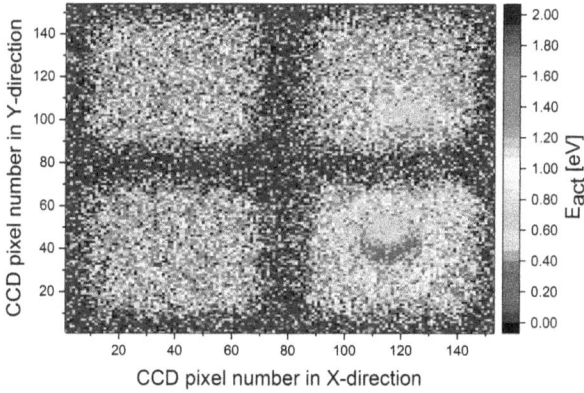

(b) Map of E_{act}

Figure 6.20: (a) Light emission image of a fraction of the PM3350T STD at $\Delta V = 4$ V, $t_{exp} = 1$ h and $T = 30$ °C. (b) Map of activation energies determined from light emission images at $T_1 = 30$ °C and $T_2 = 25$ °C. One CCD pixel corresponds to 0.65 μm.

Figure 6.20(b) shows the activation energy map which corresponds to the pixel-amplitude map of figure 6.20(a). The analysis was performed at $T_1 = 30\,^{\circ}\text{C}$ and $T_2 = 25\,^{\circ}\text{C}$. A distinction between the active regions and the periphery is evident. The hotspot free regions show an activation energy of $E_{act} \gtrsim 1$ eV, which is in agreement with the dominating hole diffusion current for the PM3350T STD in this temperature range (see chapters 4 and 5). The areas attributed to hotspots, show a significantly lower $E_{act} \approx 0.5$ eV, which indicates that these areas are not dominated by $I_{diffusion}$. However, the accuracy of the determined E_{act} is rather poor. This is mainly due to the statistical nature of both the light emission and light detection process. In equation 6.9, the general expression for the variance of E_{act} is given.

$$\sigma^2\left(E_{act}\right) = \left(\frac{1}{kT_2} - \frac{1}{kT_2}\right)^{-2} \cdot \left[\left(\frac{\sigma(I_1)}{I_1}\right)^2 + \left(\frac{\sigma(I_2)}{I_2}\right)^2\right] \tag{6.9}$$

By considering the photon emission and photon detection mechanisms as processes which are described by the Poisson distribution, the variance $\sigma^2(I)$ can be expressed as shown in equation 6.10. Here, ENF_{CCD} is the excess noise factor of the CCD camera and σ_{ext}^2 is the additional variance of external noise sources, like cosmic ray peaks. Considering that $ENF_{CCD} \approx 1$ [96], equation 6.9 can be simplified to 6.11 under the condition that $\sigma_{ext} << I_{1,2}$.

$$\sigma^2(I) = I \cdot ENF_{CCD} + \sigma_{ext}^2 \tag{6.10}$$

$$\sigma\left(E_{act}\right) \approx \left(\frac{1}{kT_2} - \frac{1}{kT_2}\right)^{-1} \sqrt{\left(\frac{1}{I_1} + \frac{1}{I_2}\right)} \tag{6.11}$$

According to equations 6.9 and 6.11, $\sigma(E_{act})$ can be decreased by two approaches. The first approach is to increase the exposure time by factor n. This will result in a decrease of $\sigma(E_{act})$ by factor $\sqrt{n}$. The second approach is to average the light intensity of m micro-cells. Averaging m measurements of one random variable reduces its variance by factor m. Consequently, $\sigma(E_{act})$ will be reduced by factor $\sqrt{m}$, under the condition of invariant light emission characteristics of different micro-cells. Figures 6.21(a) and 6.21(b) show a comparison of the light emission images of a single micro-cell and an average of 192 micro-cells without hotspots at $T = 30\,^{\circ}\text{C}$ and $\Delta V = 4$ V. In figure 6.22(a), the corresponding pixel amplitudes along the $X = 40$ coordinate are shown. A significant reduction of the noise level for the averaged micro-cell is evident in both representations.

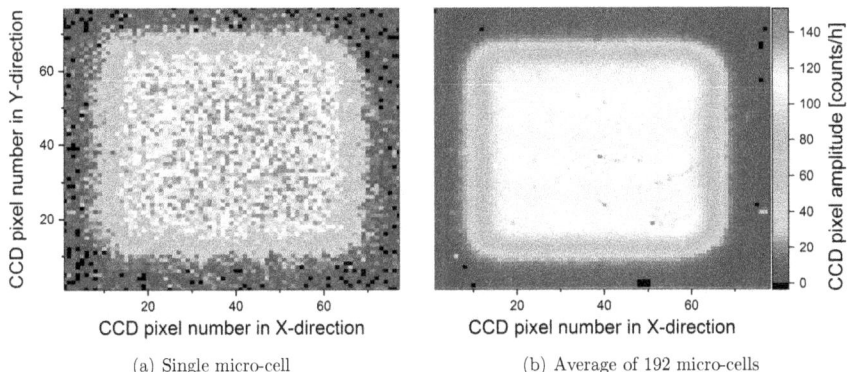

(a) Single micro-cell

(b) Average of 192 micro-cells

Figure 6.21: Light emission images of the PM3350T STD at $\Delta V = 4$ V, $t_{exp} = 1$ h and $T = 30$ °C. (a) Single micro-cell. (b) Average of 192 micro-cells. One CCD pixel corresponds to 0.65 μm.

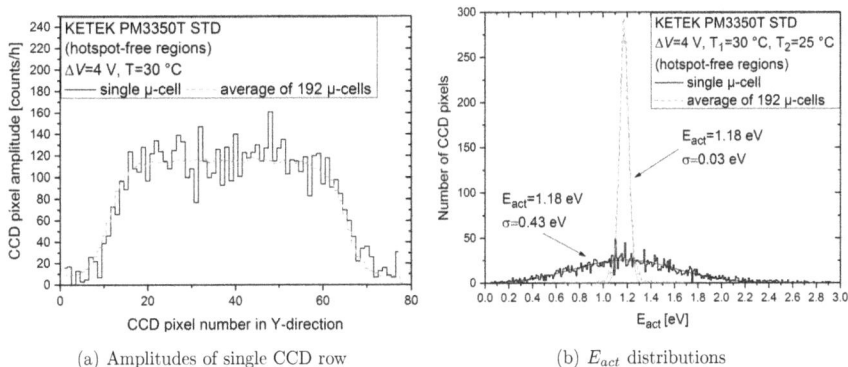

(a) Amplitudes of single CCD row

(b) E_{act} distributions

Figure 6.22: (a) CCD pixel amlitudes along the center of a single micro-cell (solid line) and an averaged micro-cell (dashed line). (b) Distributions of E_{act} within the active areas of a single micro-cell (solid line) and an averaged micro-cell (dashed line).

The activation energy was determined for every CCD pixel within the active area of the single and average micro-cell, respectively. Analogous to figure 6.20(b), the analysis was performed for $T_1 = 30$ °C and $T_2 = 25$ °C. In figure 6.22(b) the respective activation energy distributions are shown. The mean values of both distributions are located at $E_{act} = 1.18$ eV. This is in agreement with the dominating diffusion current for the PM3350T STD in this

113

temperature region, as discussed in chapters 4 and 5. The deviation from $E_{act} = 1.12$ eV is attributed to the missing $T^{3.2}$ term in equation 6.7 in contrast to equation 2.37. As expected from the previous explanations, the averaging of 192 micro-cells results in a reduction of $\sigma(E_{act})$ by a factor of $\sqrt{192}$ from 0.43 eV to 0.03 eV. To achieve an equivalent accuracy by increasing t_{exp} would require an exposure time of 192 h. Besides that such long exposure times are experimentally not practicable, an additional massive deterioration of the CCD signal is expected due cosmic ray peaks.

The averaging approach is not applicable to hotspots, since they are located at different coordinates within the active area of micro-cells. However, since the light emission intensity of hotspots is 1 to 2 orders of magnitude higher than of regions without hotspots, the accuracy of the determined E_{act} is already increased.

In the following, the level to which the substrate potential suppresses the diffusion current is discussed for the PM3350T STD-SC. The utilized SiPM was fabricated with a permanent substrate potential on chip level by short circuit the p-layer and the substrate. The PM3350T STD-SC is a prototype which should demonstrate the applicability of the implemented DCR reduction and provide necessary information on possible error sources for a later series production. Besides the mentioned feature, the general structure of this prototype does not differ from the PM3350T STD-SP.

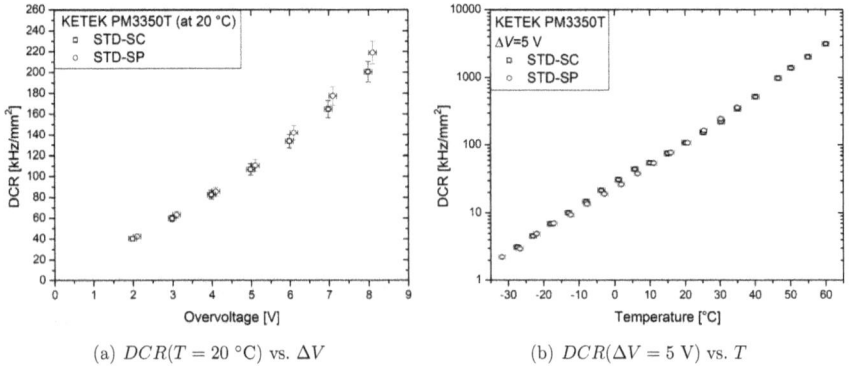

(a) $DCR(T = 20\ °C)$ vs. ΔV (b) $DCR(\Delta V = 5\ V)$ vs. T

Figure 6.23: Comparison of the dark count rates of the PM3350T STD-SC and the PM3350T STD-SP.

The experimental setup which is described in section 6.1 was used for the measurements. The covered temperature range extends from $-30\ °C$ to $+60\ °C$. In figures 6.23(a) and

6.23(b), the dark count rates of the PM3350T STD-SC and the PM3350T STD-SP are compared. No differences are observed within the uncertainties. This result demonstrates the effective implementation of the substrate potential on chip level. The Arrhenius plot in figure 6.24 corresponds to the dark count rate of the PM3350T STD-SC. Equation 4.1 with $\gamma = 2.1$ was fitted to the data points for temperatures $-30\ ^\circ\mathrm{C} \leq T \leq 30\ ^\circ\mathrm{C}$. The extracted activation energy in this temperature range amounts to $E_{act} = (0.42 \pm 0.01)$ eV. Considering the results obtained in chapter 4, it is concluded that the dark count rate of the PM3350T STD-SC is dominated by the SRH-Generation which is enhanced by field-assisted effects for $T \leq 30\ ^\circ\mathrm{C}$. For temperatures $T > 30\ ^\circ\mathrm{C}$ a faster increase of DCR is observed. This effect is attributed to a still existing diffusion current which start to dominate at higher temperatures.

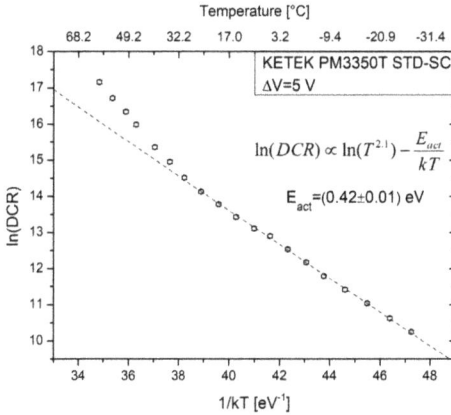

Figure 6.24: Arrhenius plot of the dark count rate of the PM3350T STD-SC.

In figure 6.25(a), a light emission image of the investigated SiPM is shown. The temperature dependence of hotspots and hotspot-free regions in the marked area was analyzed in detail. In figure 6.25(b), the corresponding regions of interest are shown with a resolution of 0.65 μm per CCD pixel. This resolution was used for the analysis.

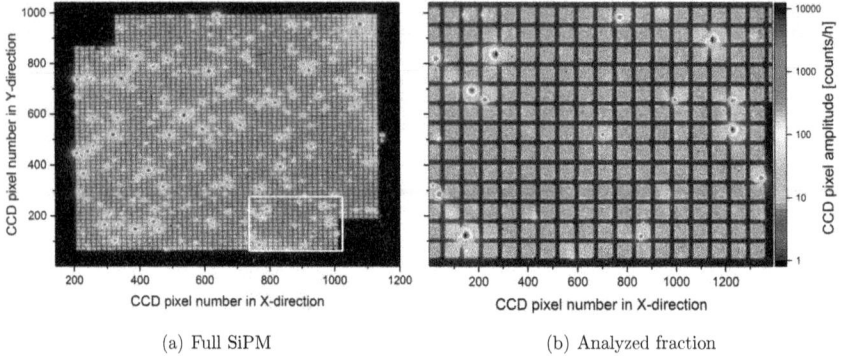

(a) Full SiPM

(b) Analyzed fraction

Figure 6.25: Light emission images of the PM3350T STD-SC at $\Delta V = 5$ V and $T = 30$ °C. (a) Full SiPM with a resolution of 3.23 μm per CCD pixel. (b) Analyzed fraction with a resolution of 0.65 μm.

Figure 6.26(a) shows the distribution of E_{act} for micro-cells without hotspots at different temperatures. The data points were described by a Gaussian distribution. The variance $\sigma^2(E_{act})$ decreases with increasing T due to the increasing light emission intensity, as discussed in the previous section. The centers of the distributions shift from (0.75 ± 0.12) eV for $T > 15$ °C to (1.13 ± 0.02) eV for $T > 55$ °C. In figure 6.26(b), the increase of E_{act} with T is shown.

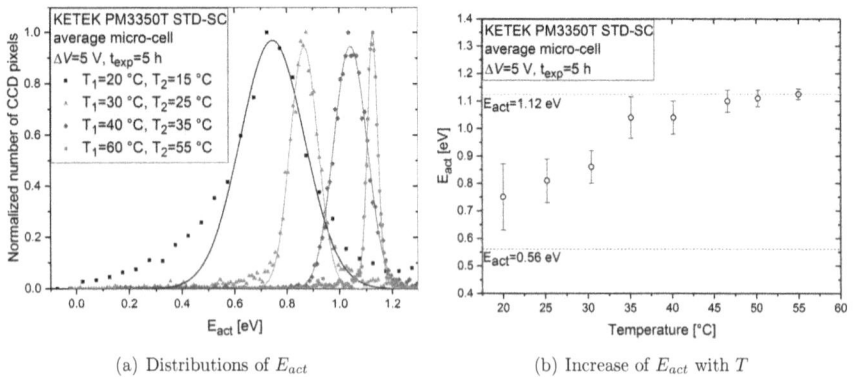

(a) Distributions of E_{act}

(b) Increase of E_{act} with T

Figure 6.26: Activation energies for micro-cells without hotspots.

Figure 6.27 shows the Arrhenius plot of the total emission intensity of the average micro-cell (without hotspots). In the temperature range 15 °C $\leq T \leq$ 30 °C, the average activation energy amounts to $E_{act} = (0.84 \pm 0.01)$ eV. For temperatures $T > 30$ °C, an average activation energy of $E_{act} = (1.10 \pm 0.02)$ eV was extracted. This is in agreement with the results presented in figure 6.26(b).

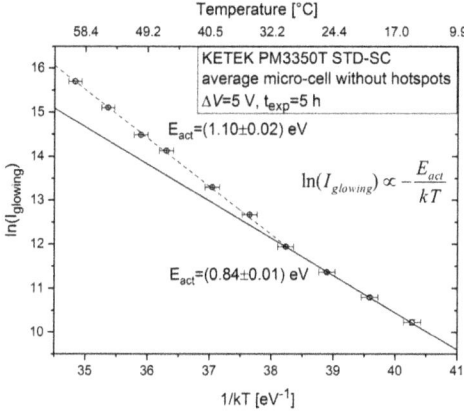

Figure 6.27: Arrhenius plot of averaged $I_{glowing}$ for hotspot-free micro-cells of the PM3350T STD-SC at $\Delta V = 5$V.

The presented analysis confirms a remaining, non-negligible contribution of the hole diffusion current to the dark count rate, despite the applied substrate potential. Additionally to the holes originating in the p-substrate, a diffusion current of minority holes from the n-layer is anticipated, as discussed in section 5.1. The application of the substrate potential does not achieve a complete depletion of the n-layer since the dopants concentration in the p-substrate is approximately 5 orders of magnitude lower than in the n-layer. One possibility to achieve a further suppression of the hole diffusion current, is to increase the dopants concentration in the n-layer and consequently to reduce the minority holes concentration, according to the mass action law ($n \cdot p = n_i^2$). However, an approach different from the increasing of the implantation dose has to be used, as will be discussed in chapter 7.

In section 6.5, the invariance of hotspots under the application of V_{substr} was discussed. In order to identify the mechanism which is responsible for the occurrence of hotspots, the temperature dependence of the light emission intensity of 5 randomly selected spots from figure 6.25(b) was investigated. For this purpose, $I_{hotspots}^{single}$ of the respective hotspot was determined by using the method described in section 6.3.2 and by additionally subtracting the average contribution from $I_{glowing}$ in the respective area. The additional correction was applied in order to further increase the accuracy of the hotspot analysis.

In figure 6.28, the Arrhenius plot of $I_{hotspots}^{single}$ normalized to $I_{hotspots}^{single}(T = 60 \,°C)$ is shown. The data points represent the average value obtained from the analysis of 5 hotspots. The function 4.1 was fitted to the data points by choosing $\gamma = 2.1$. The resulting activation energy amounts to $E_{act} = (0.51 \pm 0.01)$ eV, which is close to, but significantly lower than the mid-bandgap energy of silicon. Considering the results obtained in chapter 4, the mechanism responsible for the occurrence of hotspots is identified to be a combination of the SRH-Generation and trap-assisted tunneling. Consequently, it is concluded that the locally enhanced DCR in certain regions of the active area is caused by a spatially confined increase of the generation-recombination center density or the existence of crystal defect types with an enhanced charge carrier generation rate.

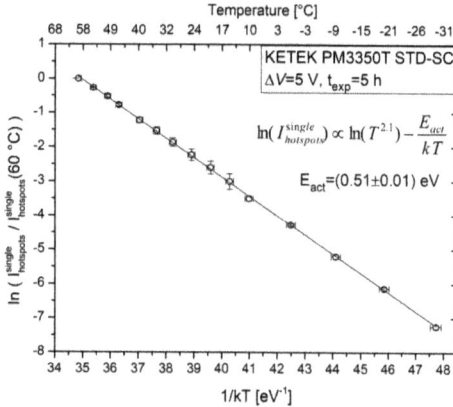

Figure 6.28: Arrhenius plot of normalized $I_{hotspots}^{single}$. The data points represent the average value from 5 hotspots of the PM3350T STD-SC.

6.7 Conclusion

In this section, the effect of hot carrier luminescence was used in order to achieve a spatially resolved measurement of the dark count rate. Within the active area of micro-cells, regions with an enhanced DCR were identified (hotspots). An novel algorithm for the determination of the number and contribution of hotspots to the overall dark count rate was presented.

The characterization of the PM3350T STD-SP showed that about 60% of the overall DCR is determined by approximately 100 hotspots.

To characterize the physical mechanisms which are responsible for dark pulses in hotspot regions and hotspot-free regions, a method to determine the spatially resolved activation energy was presented.

The responsible mechanisms for the occurrence of hotspots was identified to be the SRH-Generation with an additional contribution from field-assisted effects. The majority of generation-recombination centers is expected to be introduced to the silicon crystal during the formation of the doping profile by the implantation of ionized phosphorus and boron atoms. For this reason, the optimization of this fabrication step is mandatory to achieve a low hotspot density and consequently a low dark count rate.

Further, a prototype SiPM with an implemented substrate potential on chip level (PM3350T STD-SC) was presented. The temperature dependence of its dark count rate demonstrated the significant suppression of the diffusion current for temperatures $T \leq 30\,°C$. However, it was shown that the diffusion current is not completely suppressed in the hotspot-free areas of the PM3350T STD-SC and may become dominant for the case of a reduced hotspot density.

Chapter 7

Optimization of Ion Implantation Parameters

In chapter 6, the contribution of hotspots to the dark count rate was discussed. The origin of hotspots was attributed to crystal defects generated during the ion implantation of the p- and n-layers. This chapter addresses the optimization of the n-layer implantation with regard to avoiding crystal defects inside the active region. In section 7.1, the simulated distribution of crystal defects ,which are generated by the ion implantation, is discussed for two implantation energies.

Sections 7.2 - 7.4 address the optimization of the n-layer implantation based on the simulation results. The impact of the ion implantation energy, dose and depth on the hotspot density and on the contribution of hotspots to the total dark count rate is reviewed.

7.1 Simulation of implantation defects with SRIM

In order to study the crystal defects which are generated by the implantation of ionized phosphorus atoms, the software SRIM (**S**topping and **R**ange of **I**ons in **M**atter)[99] was used in this work. SRIM is a Monte-Carlo simulator which performs calculations of stopping powers, range and straggling distributions for ions of a certain energy in elemental and multi-layer targets. In this work, phosphorus implantation energies (E_{Imp}^P) of 3.5 MeV and 4.5 MeV were studied. The simulated implantation profiles are shown in figure 7.1. The multi-layer target consists of 1 μm SiO_2 on top of elemental Si as used in the fabrication process. As expected, the ions with the higher energy are stopped at a larger depth within the silicon crystal. The peak of the ion distribution is shifted by 400 nm moving from 3.5 MeV to 4.5 MeV. The results obtained from this simulation are consistent with performed SIMS (Secondary Ion Mass Spectroscopy) measurements of the concentration profile of phosphorus implanted with an energy of 3.5 MeV (see figure 7.2).

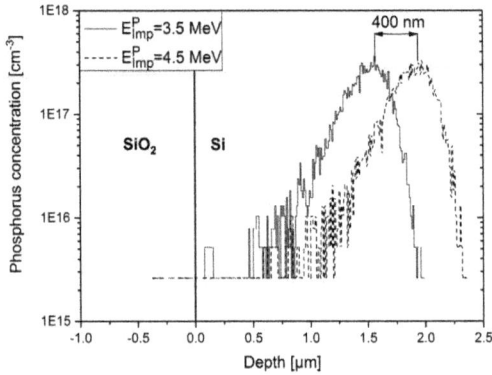

Figure 7.1: Simulated implantation profiles of phosphorus with energies of 3.5 MeV and 4.5 MeV.

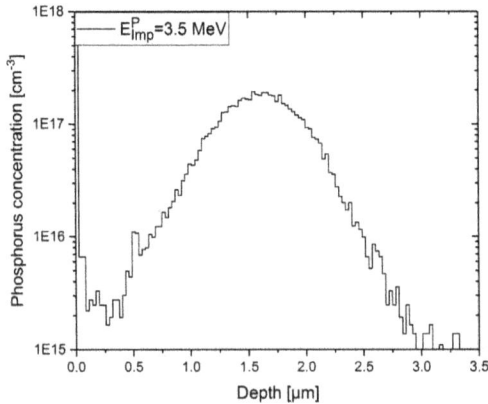

Figure 7.2: SIMS measurement of the ion implantation profile of phosphorus implanted with an energy of 3.5 MeV.

The SIMS measurements were performed after an annealing step, which explains the broader phosphorus distribution with respect to the simulated results. Nevertheless, the peak of the phosphorus concentration is located at about 1.6 μm within the silicon for both the simulated and the measured profiles.

The utilized software also provides the vacancy distribution which is generated by the impinging ions during implantation. These vacancies are equivalent to a displacement of a silicon atom from its original lattice position (point defect). An agglomeration of point defects, depending on the fluence, particle type and energy, results in disordered regions that are referred to as cluster defects [100]. Cluster defects are formed at the end of the recoil paths and show an enhanced charge carrier generation rate, which leads to an increased I_{dark} [101]. However, a description of defect clustering is not provided by SRIM. In this work, it is assumed that the density of all kinds of generated crystal defects is proportional to the simulated density of vacancies.

In figure 7.3, the simulated vacancy concentrations for implantation energies of $E_{Imp}^{P} = 3.5$ MeV and $E_{Imp}^{P} = 4.5$ MeV are shown. As expected, the total number of introduced vacancies increases with the ion energy. For 3.5 MeV, 5218 vacancies per ion are generated, whereas phosphorus with an energy of 4.5 MeV introduces 5456 vacancies per ion. Defects that are located at the maximum electric field and hence at the maximum P_{trigg} are contributing most to the dark count rate of SiPMs. For the investigated SiPMs, the depletion region is expected to extend to the peak of the phosphorus distribution. The maximum electric field is expected to be located at a depth of roughly 0.6 μm in the silicon. This is the optimum for the detection of photons with a wavelength of $\lambda = 420$ nm. In this region, the simulated vacancy distribution for $E_{Imp}^{P} = 4.5$ MeV shows a lower concentration. Consequently, the number of crystal defects which contribute to the dark count rate is expected to be lower for the $E_{Imp}^{P} = 4.5$ MeV implantation, despite the increase in the total number of introduced defects. In the fabrication process of the SiPMs, the ion implantation is followed by an annealing step. This step is not accounted for in the presented simulations. For this reason, only a qualitative prediction can be made at this point.

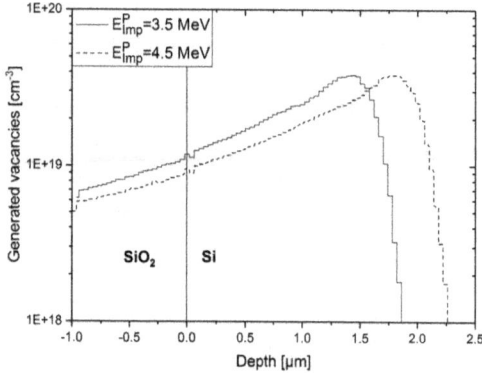

Figure 7.3: Simulated vacancy distributions generated by the implantation of phosphorus with energies of 3.5 MeV and 4.5 MeV.

7.2 Increase of the ion implantation energy

The simulation results discussed in section 7.1 predict a lower density of crystal defects in the multiplication region for higher phosphorus implantation energies E_{Imp}^P. In this section, an experimental study of the impact of E_{Imp}^P on the number of hotpots and on their contribution to the dark count rate is presented. The energy of the implanted phosphorus which forms the n-layer, was modified for the PM3350T from the standard 3.5 MeV to 4.5 MeV. In table 7.1, the key features of the investigated samples are summarized. For both SiPM types, the substrate potential was applied by using the method described in section 5.1. The low-light-level investigations were performed by using the method described in chapter 6.

In figure 7.4(a), the IV-characteristics of both SiPM types are shown under reverse bias conditions. With the increase of the n-layer implantation energy, the breakdown voltage increased from (25.6 ± 0.1) V to (34.6 ± 0.1) V. This effect is due to the increased depth of the pn-junction. The electric field increases more slowly with the applied voltage for the PM3350T HE-SP and hence the impact ionization threshold is reached at higher voltages with respect to the PM3350T STD-SP. Additionally, the slope with which the breakdown voltage decreases with temperature changed from about 20 mV/K for the PM3350T STD-SP to about 25 mV/K for the PM3350T HE-SP. In figure 7.4(b), box plots of the dark current of both SiPM types are shown at a fixed overvoltage of $\Delta V = 4$ V. For the PM3350T HE-SP, I_{dark} is significantly reduced. This result was expected, since the device capacity and consequently the gain is reduced due to the increased width of the depletion region.

Table 7.1: List of investigated samples - Implantation energy

SiPM type	PM3350T STD-SP	PM3350T HE-SP
E_{Imp}^{P} [MeV]	3.5 MeV	4.5 MeV
Impl. dose [cm^{-2}]	$1 \cdot 10^{13}$ cm^{-2}	$1 \cdot 10^{13}$ cm^{-2}
V_{BD} [V]	(25.6 ± 0.1) V	(34.6 ± 0.1) V
Substrate potential	Yes	Yes
N_{cells}	3600	3600
Pitch [μm]	50 μm	50 μm
Number of samples	6	6

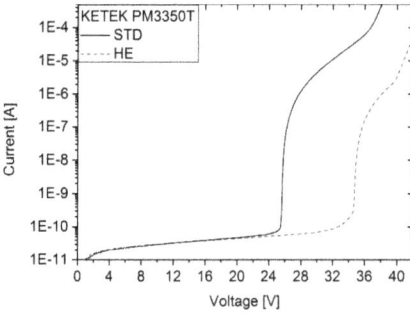

(a) IV-characteristics

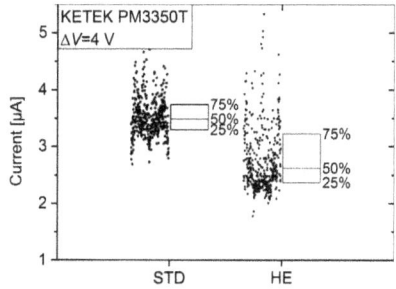

(b) Box plot at $\Delta V = 4$ V

Figure 7.4: Current-voltage characteristics of the PM3350T STD and the PM3350T HE (without substrate potential) at dark conditions.

The photon detection efficiency is a crucial parameter for the application of SiPMs. Therefore the comparison of different detector technologies at a fixed absolute or relative PDE is the most reasonable way. However, the determination of PDE is not always accessible and requires a special setup. For this reason, in literature the SiPMs are often compared at fixed absolute overvoltages $\Delta V = (V - V_{BD})$ or relative overvoltages $\Delta V_{rel} = (V - V_{BD})/V_{BD}$. In figure 7.5, the detection efficiency of the PM3350T STD-SP and the PM3350T HE-SP are compared at $\lambda = 406$ nm, using the method described in section 3.5. No significant differences in $PDE(\Delta V)$ are be observed between both technologies. For this reason, the comparison at a fixed PDE and a fixed overvoltage is equivalent. In the following, the

7. Optimization of Ion Implantation Parameters

detectors are compared at a fixed absolute overvoltage, to be consistent and comparable with measurements reported in literature.

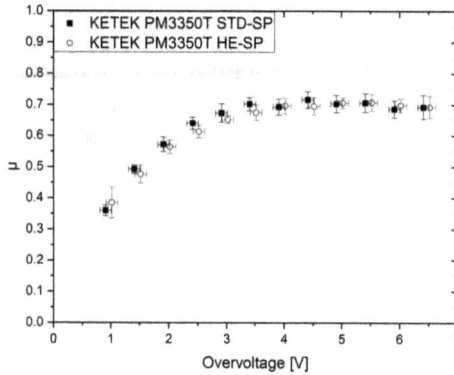

Figure 7.5: Comparison of the photon detection efficiency of the PM3350T STD-SP and the PM3350T HE-SP at $\lambda = 406$ nm.

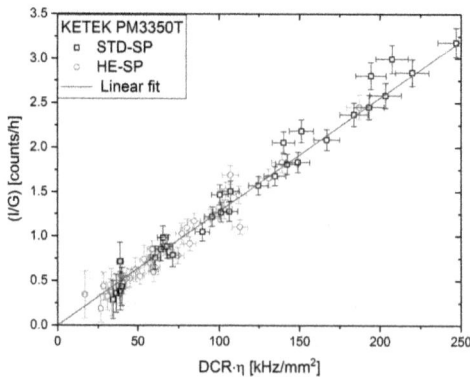

Figure 7.6: Linear correlation of I/G and $DCR \cdot \eta$. The data points correspond to the KETEK PM3350T STD-SP and HE-SP.

Analogous to figure 6.12, figure 7.6 shows the quantity I/G plotted versus $DCR \cdot \eta$ for both SiPM types. Despite a 25 % difference in the gain of both technologies, they show the same dependence of I/G on $DCR \cdot \eta$. This result indicates that the emitted wavelength spectrum is similar, which allows for a direct comparison of both SiPM types using the method presented in chapter 6. In figure 7.7, the average number of detected hotspots is shown as a function of ΔV for both device types respectively. For the PM3350T HE-SP, $N_{hotspots}$ at $\Delta V = 4$ V is reduced by a factor of approximately 2 with respect to the PM3350T STD-SP. Since the operation range of both SiPMs differ (see figure 7.4(a)), the PM3350T HE-SP was only investigated at $\Delta V \leq 6.4$ V. For higher overvoltages, a rapid decrease of $N_{hotspots}$ was observed due to effects described in section 6.4.

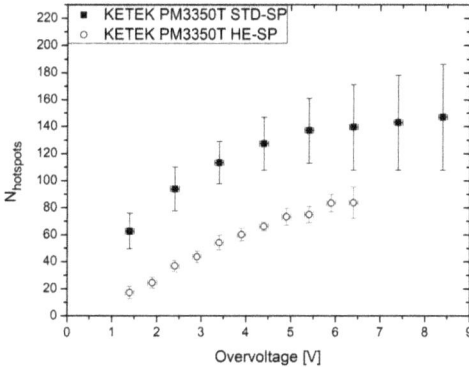

Figure 7.7: Number of hotspots versus the overvoltage for the KETEK PM3350T STD-SP and the PM3350T HE-SP.

In order to compare the average light intensities emitted by hotspots, $I_{hotspots}$ is normalized to the product of G, $N_{hotspots}$ and t_{exp}. The resulting quantity describes the amount of counts produced in the CCD per electron-hole pair in the avalanche breakdown, per hour and per hotspot. This method is valid, since the energy spectra of the emitted photons are comparable.

From figure 7.8 it is evident that at $\Delta V = 4$ V the normalized $I_{hotspots}$ of the PM3350T HE-SP is reduced by approximately a factor of 2 with respect to the PM3350T STD-SP. This leads to the conclusion that in addition to the reduced number of contributing crystal defects, the dark count rate of these defects is also reduced by an increased implantation energy. This result is in a qualitative agreement with the SRIM simulations. The crystal defects are generated deeper inside the silicon when using a higher E_{Imp}^P. The location of the maximum electric field is expected to remain at approximately the same depth as the implantation energy is increased, since the boron implantation profile for the p-layer is not changed. For this reason, the effective P_{trigg} at the coordinates of the average hotspot is reduced. This ansatz simultaneously explains the reduced $N_{hotspots}$ and the reduced DCR of hotspots. Considering both effects, the increase of E_{Imp}^P from 3.5 MeV to 4.5 MeV reduced the dark count rate generated by hotspots by approximately a factor of 3 to 4.

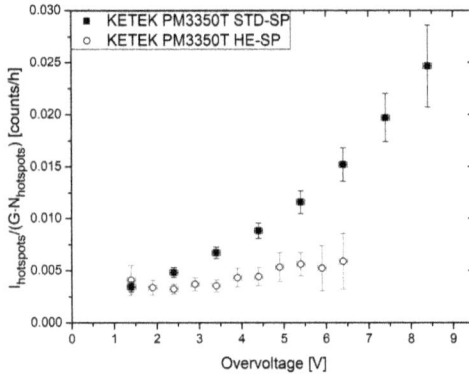

Figure 7.8: I/G of the KETEK PM3350T STD-SP and the PM3350T HE-SP, normalized to the total number of detected hotspots and the exposure time.

In figure 7.9, the contribution of hotspots to the total dark count rate is shown. The determined impact is significantly reduced from approximately 55 % for the PM3350T STD-SP to approximately 30 % for the PM3350T HE-SP. At higher overvoltages, R of the PM3350T HE-SP is decreasing. Analogous to $N_{hotspots}$, the reason for this effect is the transition from the operation range to the non-operation range, as discussed in section 6.4.

In figures 7.10(a) and 7.10(b), examples of light emission images at $\Delta V = 5.4$ V are shown for the PM3350T STD-SP and the PM3350T HE-SP. The achieved reduction in the number and intensity of hotspots is clearly evident in these images.

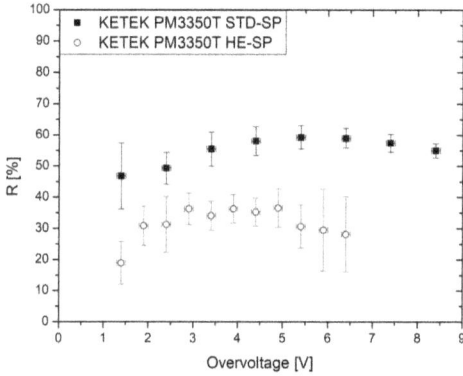

Figure 7.9: Contribution of hotspots to the total dark count rate of the PM3350T STD-SP and the PM3350T HE-SP.

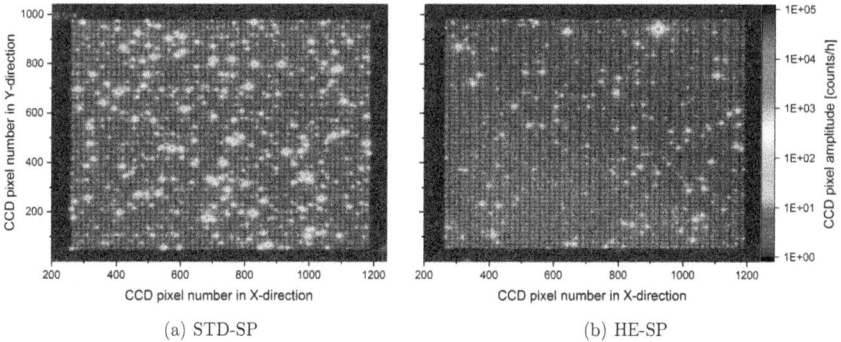

(a) STD-SP   (b) HE-SP

Figure 7.10: Light emission image of (a) the PM3350T STD-SP and of (b) the PM3350T HE-SP. The data was acquired at $\Delta V = 5.4$ V and $t_{exp} = 1$ h.

The influence of the increased implantation energy on the total dark count rate is shown in figure 7.11. Due to the significant reduction of R, DCR at $\Delta V = 4$ V is reduced from about 105 khz/mm^2 for the PM3350T STD-SP to about 50 khz/mm^2 for the PM3350T HE-SP at (21 ± 1) °C.

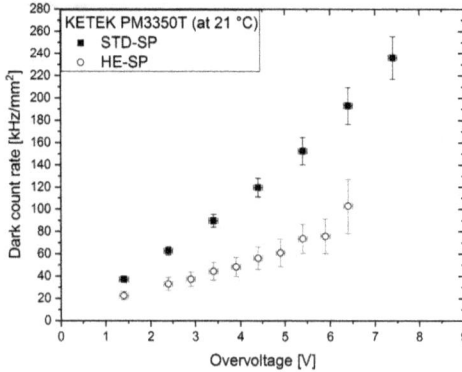

Figure 7.11: Dark count rate at 21 °C as a function of overvoltage for the KETEK PM3350T STD-SP and HE-SP.

In figure 7.12(a), the dark count rate of one sample from each SiPM technology is shown respectively as a function of temperature at $\Delta V = 4$ V. The corresponding Arrhenius plots are depicted in figure 7.12(b). The activation energies were extracted from the slopes of linear fits to the data points. For temperatures $T \lesssim -10$ °C the determined activation energies of both SiPM types match within the uncertainties and amount to approximately 0.46 eV. This is in agreement with the results presented in the previous chapters. A significant difference is observed with regard to the temperature at which the activation energy clearly deviates from 0.46 eV. For the PM3350T STD-SP, this is the case for $T > 6$ °C, whereas for the PM3350T HE-SP an increased activation energy is already observed for $T > 0$ °C. This effect is attributed to the lower hotspot density of the PM3350T HE-SP. Consequently, the remaining diffusion current starts to contribute noticeably at lower temperatures.

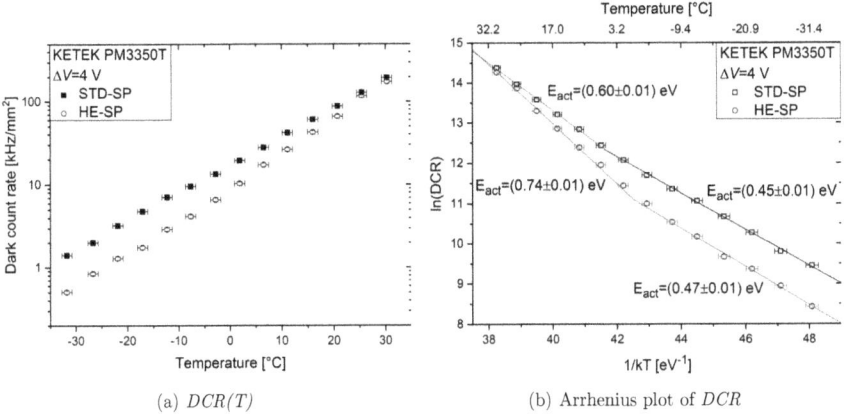

(a) $DCR(T)$ (b) Arrhenius plot of DCR

Figure 7.12: Temperature dependence of the dark count rate of the PM3350T STD-SP and the PM3350T HE-SP.

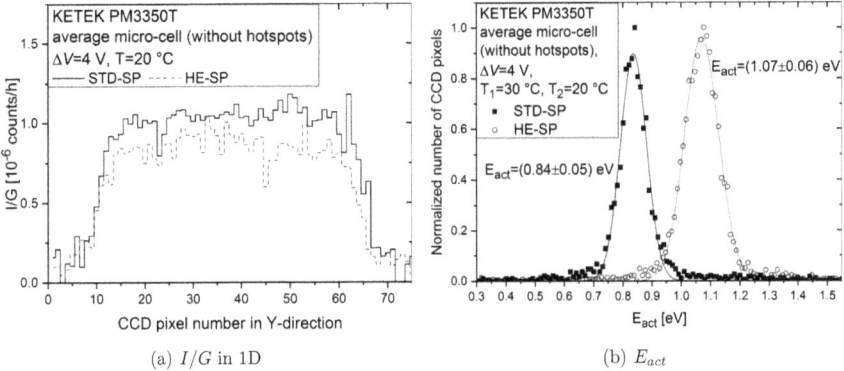

(a) I/G in 1D   (b) E_{act}

Figure 7.13: (a) Normalized light emission intensity (I/G) along the center of an average micro-cell. One CCD pixel corresponds to 0.65 μm. (b) Distribution of E_{act} for an average micro-cell without hotspots.

Using the method described in section 6.6, hotspot-free micro-cells of the PM3350T STD-SP and the PM3350T HE-SP were investigated. The measurements were carried out at $\Delta V = 4$, $t_{exp} = 5$ h, $T_1 = 30\,°C$ and $T_2 = 20\,°C$. In figure 7.13(a), the normalized light emission intensity I/G along the center of the average micro-cell is shown for both

131

SiPM types respectively. The PM3350T HE-SP shows a lower I/G. This result is in a qualitative agreement with the lower DCR due to a reduced density of crystal defects that are introduced by the phosphorus implantation. In figure 7.13(b), the activation energies of hotspot-free areas of both SiPMs are shown. An increase from $E_{act} = (0.84 \pm 0.05)$ eV for the PM3350T STD-SP to $E_{act} = (1.07 \pm 0.06)$ eV for the PM3350T HE-SP is observed. This is in agreement with the Arrhenius plot shown in figure 7.12(b) and confirms the still existing contribution of $I_{diffusion}$ for both SiPM types.

7.3   Decrease of the ion implantation dose

In the previous section, the achieved reduction of the dark count rate by the increase of the phosphorus implantation energy was discussed. In this section, a second approach aiming for the suppression of crystal defects is presented.

The density of generated defects is expected to be proportional to the concentration of the implanted phosphorus ions. Consequently, the dose of the implanted dopants was reduced by a factor of 2 from the standard $1 \cdot 10^{13}$ cm^{-2} to $5 \cdot 10^{12}$ cm^{-2}. The reference devices with the standard implantation parameters were produced in the same batch, in order to ensure the comparability of the SiPMs. In table 7.3, the key features of the investigated samples are summarized. Both technologies were investigated with and without the substrate potential V_{substr}.

Table 7.2: List of investigated samples - Implantation dose

SiPM type	PM4450T STD/STD-SP	PM4450T LD/LD-SP
E_{Imp}^{P} [MeV]	3.5 MeV	3.5 MeV
Impl. dose [cm^{-2}]	$1 \cdot 10^{13}$ cm^{-2}	$5 \cdot 10^{12}$ cm^{-2}
V_{BD} [V]	(27.1 ± 0.1) V	(32.6 ± 0.1) V
Substrate potential	No/Yes	No/Yes
N_{cells}	5056	5056
Pitch [μm]	50 μm	50 μm
Number of samples	2	2

In figure 7.14, the dark count rates of the PM4450T STD/STD-SP, the PM4450T LD/LD-SP and the PM3350T STD-SP are shown as a function of the overvoltage at 21 °C. The dark count rate of the PM4450T LD is by approximately a factor of 1.7 higher with respect to the PM4450T STD. This is qualitatively explained by the lower phosphorus concentration in the n-layer. As discussed in chapter 4, the KETEK SiPMs are dominated by the diffusion of minority holes from the substrate and the n-layer into the multiplication region. According to the mass action law ($n \cdot p = n_i^2$), the concentration of the free holes in the n-layer increases with decreasing phosphorus concentration. Additionally, the mean free path of holes in the n-doped region is increasing. Both effects lead to an enhanced contribution of the minority hole diffusion to the dark count rate.

The application of the substrate potential reverses the DCR-ratio of both technologies. Due to the suppression of the minority hole diffusion by the substrate potential, the contribution from crystal defects becomes the determining factor. Because the ion concentration is reduced by a factor of 2 for the PM4450T LD/LD-SP prototypes, also the contribution from hotspots is reduced. As a consequence, the dark count rate is suppressed. At $\Delta V = 4$V, a reduction from approximately 105 kHz/mm^2 for the PM4450T STD-SP to approximately 70 kHz/mm^2 for the PM4450T LD-SP is observed.

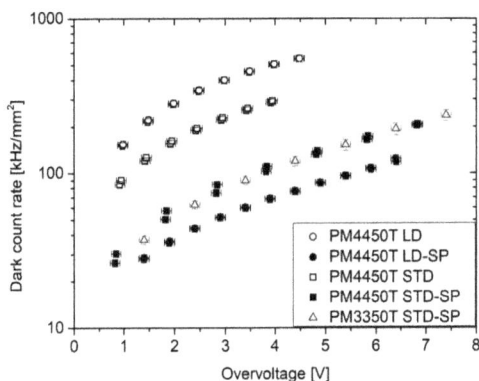

Figure 7.14: DCR at 21 °C versus the overvoltage for the PM4450T LD/LD-SP and the PM4450T STD/STD-SP.

In figure 7.15, $N_{hotspots}$ is normalized to the number of investigated micro-cells for the PM3350T STD-SP, the PM3350T HE-SP and the PM4450T LD-SP, respectively. It becomes evident, that decreasing of the implantation dose by a factor of 2, leads to a reduction of the hotspot density by approximately a factor of 1.4, which explains the reduced dark count rate. The fact that the reduction of the hotspot density is not proportional to the reduction of the phosphorus implantation dose can be qualitatively explained by the following effect:

The depletion width and hence the active region is increased with a decreasing implantation dose. This can be observed by the shift of the breakdown voltage from (27.1 ± 0.1) V (STD-SP) to (32.6 ± 0.1) V (LD-SP). Consequently, the volume in which the crystal defects contribute to the dark count rate is enlarged.

In contrast to the PM3350T HE-SP, the maximum of the phosphorus distribution is not shifted for the PM4450T LD-SP. As a result, the crytal defects are generated in a region of a larger electric field and hence a larger avalanche triggering probability. Consequently, an increased hotspot density for the PM4450T LD-SP is expected with respect to the PM3350T HE-SP, despite the lower dose of implanted phosphorus. The experimental results presented in figure 7.15 support this hypothesis. Here, the determined hotspot density for the PM4450T LD-SP is by a factor of 1.4 higher with respect to the PM3350T HE-SP.

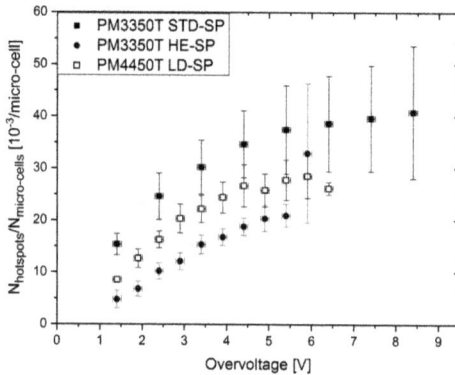

Figure 7.15: $N_{hotsptots}$ normalized with the number of investigated micro-cells for the PM3350T STD-SP, the PM3350T HE-SP and the PM4450T LD-SP.

Figure 7.16 shows box plots of the hotspot contributions to the total dark count rate. The plots contain the contributions R in the operation range of the respective SiPM types. In this range, R is independent of the overvoltage, as discussed in section 6.4. The contribution of hotspots for the PM4450T LD-SP is approximately 35 %, which is slightly increased with respect to $R \approx 30\%$ for the PM3350T HE-SP. The reason for this observation is discussed above. Assuming that the contribution from the diffusion current and high field effects for the PM3350T STD-SP and the PM4450T LD-SP is comparable, a hotspot contribution of $R \approx 0.47$ % is anticipated for the PM4450T LD-SP type. The reason for the observation of a lower value is attributed to an enhanced contribution of the minority charge carrier diffusion from the n-layer due to a lower phosphorus concentration, as discussed previously.

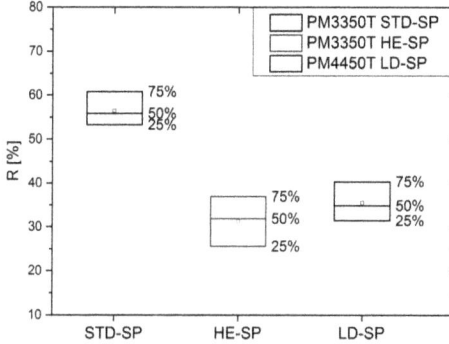

Figure 7.16: Box plots of R within the operation range of the PM3350T STD-SP, the PM3350T HE-SP and the PM4450T LD-SP.

7.4 Variation of the n-layer depth

In the previous section, the impact of an increased E_{imp}^{P} and the impact of a decreased dopants concentration on the hotspot density and hence on the dark count rate was discussed. The depth of the phosphorus concentration peak was identified to be a crucial parameter. A shift of the peak value causes a shift of the crystal defects to coordinates of a higher or a lower P_{trigg}. This has a significant impact on the number and on the dark count rate of generated hotspots.

In this section, the peak concentration of the implanted phosphorus is varied at a fixed $E_{Imp}^{P} = 4.5$ MeV and a fixed implantation dose of $1 \cdot 10^{13}$ cm^{-2}. The variation was realized by a modification of the scatter oxide thickness from the standard 1.0 μm to 1.5 μm, as shown schematically in figure 7.17. The resulting shift of the phosphorus peak concentration led to a shift of V_{BD} from (35.4 ± 0.1) V for the PM3350T HE-SP to (26.2 ± 0.1) V for the PM3350T HE-S-SP. In table 7.3, the key parameters of the compared SiPMs are shown.

7. Optimization of Ion Implantation Parameters

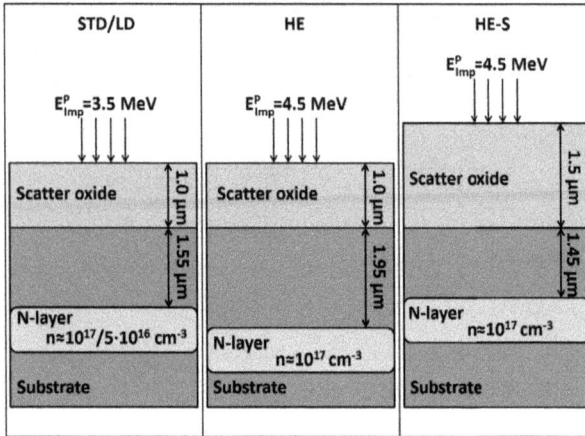

Figure 7.17: Sketch of the realized shift of the n-layer depth by a variation of the implantation energy and the scatter oxide thickness.

Table 7.3: List of investigated samples - N-layer depth

SiPM type	PM3350T HE-SP	PM3350T HE-S-SP
E_{Imp}^P [MeV]	4.5 MeV	4.5 MeV
Impl. dose [cm^{-2}]	$1 \cdot 10^{13}$ cm^{-2}	$1 \cdot 10^{13}$ cm^{-2}
Scatter oxide thickness	1000 nm	1500 nm
V_{BD} [V]	(35.4 ± 0.1) V	(26.2 ± 0.1) V
Substrate potential	Yes	Yes
N_{cells}	3568	3568
Pitch [μm]	50 μm	50 μm
Number of samples	3	3

Combining the simulation results from section 7.1 and the experimental results discussed in section 7.2, a shift of the breakdown voltage by 9 V is estimated to be caused by a 400 nm shift of the phosphorus peak concentration.

As a consequence, the dark count rate of the PM3350T HE-S-SP devices is expected to be comparable to the one of the PM3350T STD-SP. In figure 7.18, this comparison is drawn at $T = 21$ °C. The PM3350T HE-SP shows a dark count rate of approximately 40 kHz/mm^2 at $\Delta V = 4$ V. This value is 10 kHz/mm^2 lower compared with the results presented in section

136

7.2. The PM3350T HE-S-SP shows a dark count rate of approximately 95 khz/mm^2 at $\Delta V = 4$ V, which is also 10 khz/mm^2 lower compared with the PM3350T STD-SP presented in section 7.2. The observed variations from the expected values are not considered to be significant at this point due to the small number of investigated samples.

With the results obtained in this section, it is confirmed that the achieved reduction of the dark count rate in section 7.2 is due to the shift of the phosphorus peak concentration and hence the crystal defect coordinates with respect to P_{trigg}. The choice of E_{Imp}^P by itself is not expected to show a strong impact on the dark count rate.

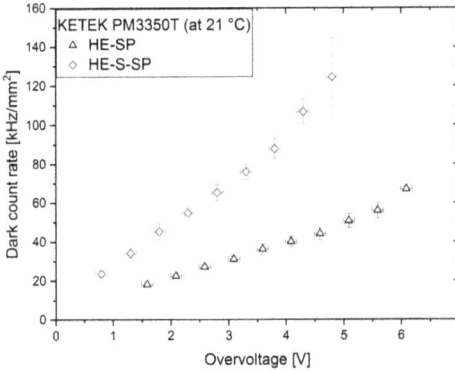

Figure 7.18: Dark count rate versus overvoltage for the PM3350T HE-SP and the PM3350T HE-S-SP, at 21 °C.

7.5 Conclusion

This chapter discussed the suppression of the crystal defects in the active region of micro-cells and the consequential reduction of the dark count rate. The improvements were realized by an optimization of the ion implantation parameters for blue sensitive KETEK SiPMs with a maximum PDE at $\lambda = 420$ nm.

It was found that one key parameter is the position of the phosphorus peak concentration of the n-layer. A shift of this peak to a larger depth reduces the defect density in the active region. The shift can be achieved either by a higher phosphorus implantation energy E_{Imp}^P or by a thinner scatter oxide. In this work, the implantation energy was increased from

3.5 MeV (PM3350T STD-SP) to 4.5 MeV (PM3350T HE-SP), which caused a shift of the phosphorus peak concentration by approximately 400 nm. As a consequence, the number and the contribution of hotspots to the dark count rate was significantly reduced. This resulted in a reduction of the total dark count rate from 105 kHz/mm^2 for the PM3350T STD-SP down to 40 kHz/mm^2 for the PM3350T HE-SP at $\Delta V = 4$ V. Analogous to the results presented in section 6.6 for the PM3350T STD-SC, a non-negligible contribution of $I_{diffusion}$ for both the PM3350T STD-SP and the PM3350T HE-SP was observed. Due to a lower number of defects in the active region, the remaining diffusion current of the PM3350T HE-SP shows a higher relative contribution with respect to the PM3350T STD-SP.

Besides the reduced crystal defect density, the electric field distribution is also influenced by the variation of the implantation profile. The relation of the reduced number of hotspots with the changed electric field was not reviewed in this thesis. Both SiPMs show an equal PDE at $\lambda = 406$ nm. From this result it can be concluded that the avalanche triggering probabilities and hence the electric fields are comparable within the first 0.5 μm. Since the detected hotspots are expected to be located within a depth of 0.5 μm (see section 6.3.2), the impact of the electric field is anticipated to be negligible.

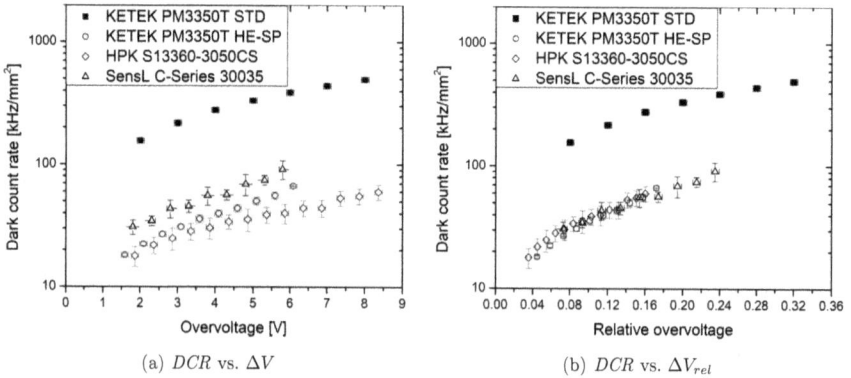

(a) DCR vs. ΔV (b) DCR vs. ΔV_{rel}

Figure 7.19: Comparison of the dark count rates of SiPMs from different producers at 21 °C.

With the combination of the substrate potential and the optimized implantation energy, the dark count rate was reduced to the level of the state of the art SiPMs, as shown in figures 7.19(a) and 7.19(b). If compared at a fixed absolute overvoltage, the dark count rate of the PM3350T HE-SP is between the values for the HPK S13360-3050CS and the SensL C-Series 30035. If the comparison is drawn at a fixed relative overvoltage, all three providers show

138

equal dark count rates. With this result, one of the major objectives of this research project was achieved.

The PM3350T HE-S-SP shows a dark count rate of 95 kHz/mm^2 at $\Delta V = 4$ V and reproduces the dark count rate of the PM3350T STD-SP with a good enough precision. For this SiPM, the shift of the phosphorus profile due to the higher implantation energy (4.5 MeV) was compensated by a thicker scatter oxide. This result confirms that the explicit dependence of the crystal defect density on the implantation energy is negligible.

Another way to achieve a lower crystal defect density in the active region is to decrease the concentration of the implanted dopants. For the PM4450T LD-SP, the phosphorus concentration was reduced by a factor of two. This resulted in a reduction of the hotspot density and consequently a reduced DCR of 70 kHz/mm^2 at $\Delta V = 4$ V. The higher dark count rate with respect to the PM3350T HE-SP was attributed to two effects. The first effect is the enhanced contribution of the hole diffusion current originating from the lower doped n-layer. The second effect is that the phosphorus peak concentration was not shifted to a larger depth.

Chapter 8

Radiation Hardness

Silicon Photomultipliers are already used or are planned to be used in a variety of high energy physics experiments [18], [102], [103]. They are also considered for the application in space environment, where they are as well operated at harsh radiation conditions and are exposed to charged high energy particles [104], [105], [106]. The major drawback of the application of SiPMs in high luminosity colliders and space environment is their degradation due to radiation damages [19], [107]. Radiation damage occurs, when energy is deposited in the active volume of the detector in the form of atomic displacement or electron ionization [101]. In this chapter, the impact of two types of radiation on the SiPM parameters is discussed. In section 8.1, the SiPM parameters are monitored as a function of the accumulated dose of ^{60}Co γ-rays. In section 8.2, the degradation of the SiPM parameters due to the irradiation with 1 MeV neutrons is discussed.

8.1 Irradiation with ^{60}Co γ-rays

In this section, the radiation damage in KETEK SiPM caused by ^{60}Co γ-rays is reviewed. The energies of the emitted γ-rays from the ^{60}Co source are 1.17 MeV and 1.33 MeV [108]. In this energy range, the dominating interaction mechanism of ^{60}Co γ-rays with silicon is the Compton scattering [109]. The energy of the scattered electron E_e is given by the difference between the energy of the incident γ-ray E_γ and the energy of the scattered γ-ray E'_γ. The well known expression for the energy of the scattered γ-ray is given by equation 8.1. Here, ϕ is the angle between the incident and the scattered γ-ray, $m_0c^2 = 511$ keV is the rest energy of the electron. The maximum energy transfer to the electron occurs in a collision with $\phi = \pi$. For $E_\gamma = 1.33$ MeV, the maximum achievable electron energy is $E_e = 1.12$ MeV. Electrons with an energy larger than 260 keV are able to dislocate silicon atoms and create point defects in the crystal [100]. To create a cluster defect, the kinetic energy of an electron must be larger than 8 MeV [110]. For this reason, the irradiation with a ^{60}Co γ-source allows for the investigation of damages caused by point defects only [111].

$$E'_\gamma = \frac{m_0 c^2}{(1 - cos(\phi) + m_0 c^2/E_\gamma)} \tag{8.1}$$

The PM1150T STD SiPMs were exposed to a ^{60}Co γ-ray source and doses of 40 Gy, 120 Gy, 160 Gy and 240 Gy were accumulated. No external bias voltage was applied during the irradiation. The irradiations were performed at the *National Institute for Laser Plasma and Radiation Physics* in Magurele, Romania, by the group of Dr. Dan Sporea. The characterization of the SiPMs was performed two weeks after the irradiation. Since the devices were stored at room temperature, a certain level of annealing is expected. In figure 8.1(a), the current-voltage characteristics of the PM1150T STD at dark conditions are shown for every dose. As a reference, a non-irradiated PM1150T STD from the same batch was used. An increase of the dark current is observed at voltages below and above the breakdown voltage. This result indicates that both surface and bulk defects are generated by the γ-rays

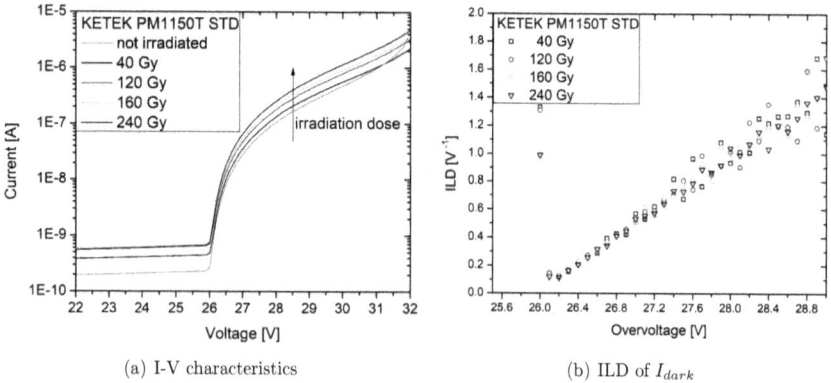

(a) I-V characteristics

(b) ILD of I_{dark}

Figure 8.1: (a) Dark current of the PM1150T at different accumulated doses of ^{60}Co γ-rays. (b) ILD of the measured dark currents.

In figure 8.1(b), the inverse logarithmic derivative (ILD) of I_{dark} (see section 3.4) is plotted versus the applied reverse bias voltage. No impact of the irradiation dose on the breakdown voltage is observed. This result was confirmed by using the relative gain for the determination of V_{BD}, which leads to the conclusion that no change in doping concentration was caused by the irradiation. In figure 8.2(a), the normalized pulse-height spectra of a non-irradiated and of an irradiated PM1150T STD-SP are shown at $\Delta V = 4$ V. The optical crosstalk probability also did not change with the irradiation dose.

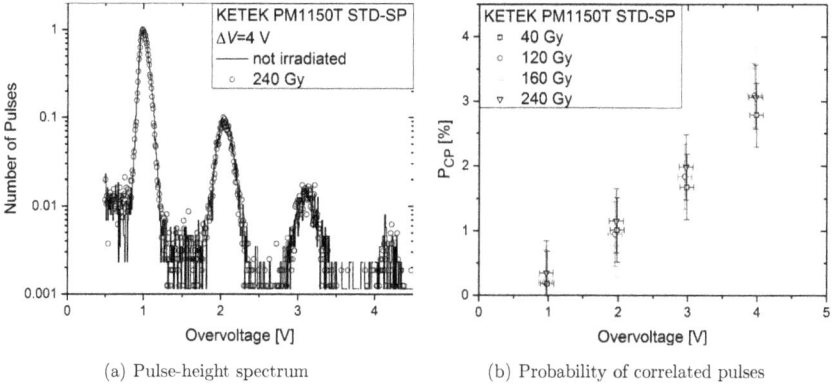

(a) Pulse-height spectrum

(b) Probability of correlated pulses

Figure 8.2: (a) Comparison of the pulse-height spectra of a not irradiated PM1150T and a PM1150T irradiated with ^{60}Co γ-rays (240 Gy). (b) Probability of correlated pulses for different accumulated ^{60}Co γ-ray doses.

In figure 8.2(b), the probability of correlated pulses is shown as a function of overvoltage for different irradiation doses. Also here, no influence of the accumulated dose is observed. This result indicates that no shallow traps were introduced by the irradiation, which would otherwise lead to an increase of the afterpulsing probability.

In figure 8.3(a), the dark count rate is plotted versus the irradiation dose at different overvoltages. A linear increase of DCR with the accumulated dose is observed, which is attributed to a linear increase of the point defect density. In figure 8.4, light emission images of the irradiated PM1150T STD-SP are shown. The increasing number of hotspots with irradiation dose is clearly evident in these images. In figure 8.3(b), the average light emission intensity from one micro-cell is plotted versus the accumulated dose. Here, 221 micro-cells were analyzed. The increasing error bars indicate the variation of the number of point defects per micro-cell. The observed linear increase of the light intensity is equivalent to the linear increase of the dark count rate.

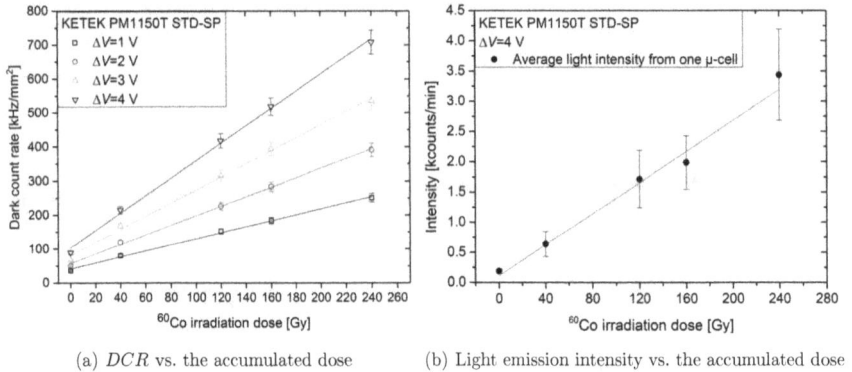

(a) DCR vs. the accumulated dose   (b) Light emission intensity vs. the accumulated dose

Figure 8.3: (a) Linear increase of (a) the dark count rate and (b) the average light intensity emitted by a micro-cell with increasing accumulated ^{60}Co γ-ray dose. The error bars in (b) represent the spread of the mean over the 221 analyzed micro-cells.

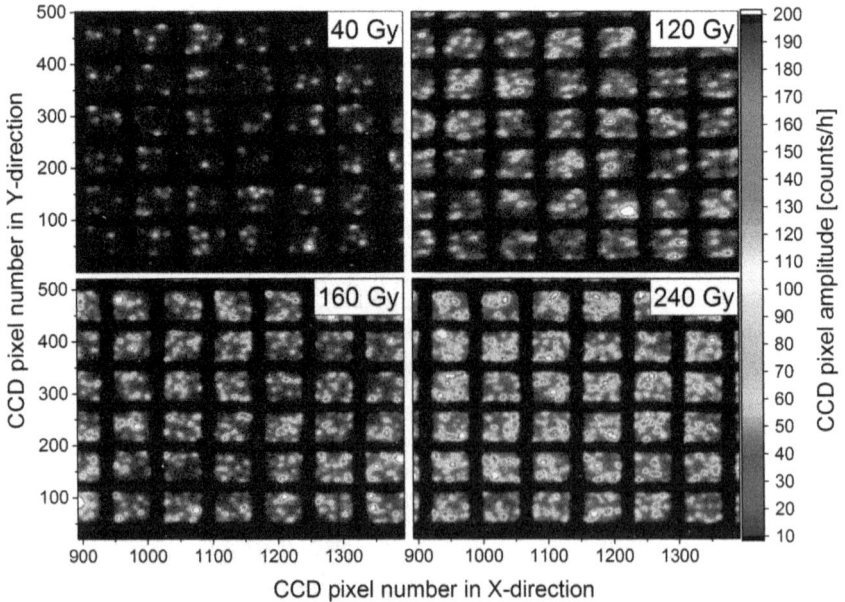

Figure 8.4: Light emission images of the PM1150T STD-SP at $\Delta V = 4$ V for different accumulated ^{60}Co γ-ray doses.

144

In order to evaluate the dark count rate which is generated by point defects, the CCD pixel amplitude distributions of the light emission images were compared. The result is shown in figure 8.5(a). The CCD pixel amplitudes of ion implantation damages are expected to reach 10 to 100 counts per minute. In rare cases, also amplitudes of about 1000 counts per minute were observed. In this amplitude range, no influence of the γ-ray irradiation was observed. However, an increase of the number of CCD pixels with an amplitude smaller than 10 counts per minute is noticeable. In figure 8.5(b), a direct comparison of a hotspot caused by an ion implantation defect and hotspots caused by γ-ray defects is shown. The conclusion is, that point defects caused by ^{60}Co γ-rays generate a dark count rate which is approximately one order of magnitude smaller than the dark count rate generated by ion-implantation defects.

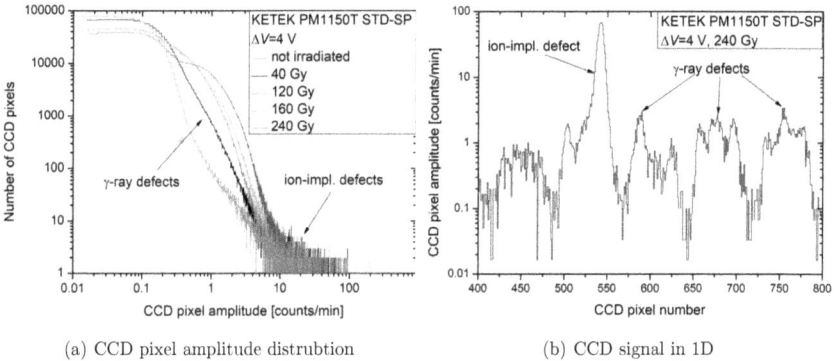

(a) CCD pixel amplitude distrubtion   (b) CCD signal in 1D

Figure 8.5: Comparison of hotspots due to ion implantation damages and point defects generated by ^{60}Co γ-rays.

In an additional experiment, three samples of the PM3350T STD-SP and the PM3350T HE-SP were exposed to the ^{60}Co source. The accumulated dose amounted to 120 Gy. In figure 8.6, the absolute increase of the dark count rate after irradiation is shown for both SiPM types. The PM3350T HE-SP shows a higher tolerance to the irradiation than the PM3350T STD-SP. At an overvoltage of $\Delta V = 4$ V, the absolute increase of the dark count rate differs by about 80 kHz/mm^2.

As discussed in section 7.2, the active region of the PM3350T HE-SP is larger than for the PM3350T STD-SP. Consequently, the number of defects which contribute to the dark count rate is expected to be larger for the PM3350T HE-SP, if the generated point defects are assumed to be uniformly distributed in the whole SiPM volume. From this point of view, a

larger increase of the dark count rate is expected for the PM3350T HE-SP with respect to
the PM3350T STD-SP. So far, the obtained result is not understood.

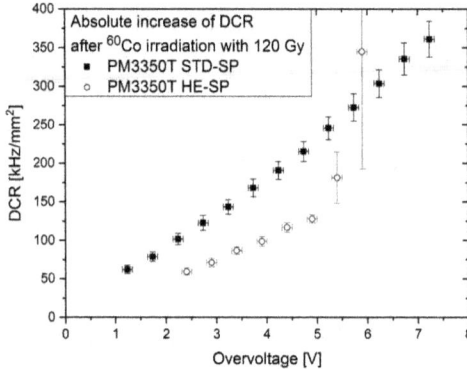

Figure 8.6: Absolute increase of the dark count rate of the PM3350T STD-SP and the
PM3350T HE-SP after the irradiation with ^{60}Co γ-rays (120 Gy). The error bars correspond
to the spread of the mean over three SiPMs.

A possible explanation is the difference of the electric field distributions and hence the difference of the position-dependent avalanche triggering probabilities of both SiPM types. To
determine the 3D-distribution of the avalanche triggering probability inside the SiPM micro-
cell, the following two methods are proposed:

(i) For the first method, a pulsed light source with adjustable wavelengths is required. One
micro-cell of the SiPM is illuminated with a focused light spot with a diameter of 1 to 2 μm.
The approach described in section 3.5 is used to determine the photon detection efficiency
in the restricted area of the light spot. By scanning the complete micro-cell area, a lateral
distribution of the avalanche triggering probability is obtained. Using different wavelengths,
a 3D-profile can be generated.

(ii) For the second method, also a light source with adjustable wavelengths is required. But
it does not need to be pulsed. Additionally, a low-light-level camera is needed. The micro-
cell area is illuminated homogeneously with a certain wavelength. The light which is emitted
by the SiPM due to the effect of hot carrier luminescence is detected with the low-light-level
camera using the setup shown in figure 6.1. The obtained light intensity distribution is
equivalent to the distribution of the avalanche triggering probability (see chapter 6). Analogous to the first method, a 3D-profile can be obtained by using different wavelengths. To

ensure that the detected light intensity is not distorted by the reflected excitation light, an adjusted wavelength filter can be mounted in front of the camera. Alternatively, an image which is recorded with the SiPM bias voltage set to zero can be subtracted from the image obtained for the biased SiPM.

As mentioned earlier, an annealing effect already sets in at room temperature. In figure 8.7(a), the measurements of the dark current of the irradiated PM3350T STD are shown. The first measurement was performed two weeks after the irradiation. The second measurement was performed nine weeks after the irradiation. The SiPMs were stored at room temperature in between the measurements. As a reference, the dark current before irradiation is shown. In figure 8.7(b), the ratios of I_{dark} after irradiation divided by I_{dark} before irradiation are presented. The first measurement shows that the dark current increased by a factor of about 2.5 at voltages below the breakdown voltage. At voltages higher than the breakdown voltage I_{dark} increased by a factor of about 2 to 2.7, depending on the voltage range. In the second measurement, the dark current almost recovered for low overvoltages and was measured to be only 5 % higher than the reference current. Close to the non-operational range, the ratio is about 1.5. This result shows that the point defects in the active region of the SiPM experienced an annealing process. However, the ratio of the dark currents at voltages below the breakdown voltage increased to about 7. This effect is currently not understood.

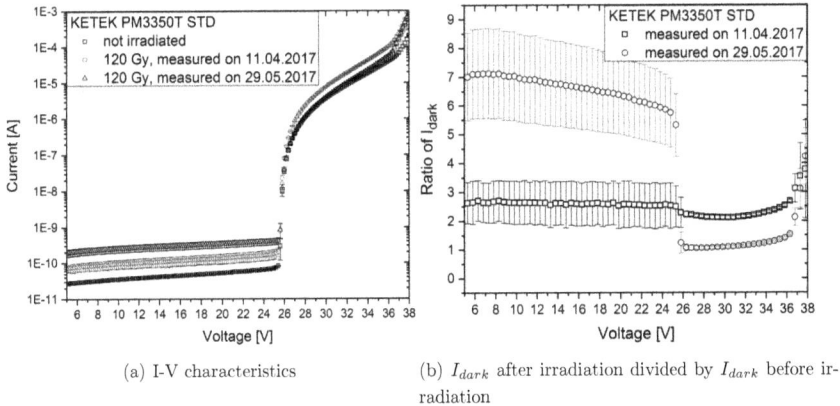

(a) I-V characteristics

(b) I_{dark} after irradiation divided by I_{dark} before irradiation

Figure 8.7: Room temperature annealing effect on I_{dark} of the irradiated PM3350T STD. The error bars in (a) correspond to the spread of the mean over three SiPMs.

8.2 Irradiation with neutrons

In this section, the impact of an irradiation with neutrons on the SiPM parameters is discussed. For the generation of a point defect, the kinetic energy of a neutron must be larger than 185 eV. If the kinetic energy exceeds approximately 35 keV, neutrons are able to generate cluster defects in the silicon crystal [112]. KETEK SiPMs were exposed to a reactor fluence of $\Phi = 10^{10}$ cm^{-2} of 1 MeV neutrons at the TRIGA Research Reactor of the JSI, Ljubljana. For this reason, the generation of cluster defects was anticipated. The irradiation was performed at room temperature without an applied bias voltage. After the irradiation, the samples were investigated by the group of Prof. Erika Garutti, at the Institute of Experimental Physics at Hamburg University [113]. In between the measurements, the SiPMs were stored in a refrigerator at -30 °C. However, during the measurements the samples were not cooled, because of which a certain level of annealing is expected. After these investigations, one 6x6 test-structure (see section 3.9) was sent cold to Munich for the characterizations which are presented in the following. As a reference, an equivalent, not irradiated device from the same batch was used.

In figure 8.8(a), the I-V measurements of the irradiated SiPM and the reference SiPM are compared. The dark current at voltages below the breakdown voltage increased by a factor of 2 to 5, depending on the voltage range. At voltages above the breakdown voltage, the dark current of the irradiated sample increased by a factor of about 400 with respect to the reference sample. This result shows that the introduced crystal damages are mostly located in the silicon bulk and not at the surface. In figure 8.8(b), the corresponding DCR measurements are shown. The investigated SiPM type was small enough to apply the standard methods for the determination of the dark count rate and the generation of pulse-height spectra up to overvoltages of $\Delta V = 3.6$ V.

In figure 8.9(a), the pulse-height spectra of the irradiated and the reference sample are shown at $\Delta V = 3.6$ V. The position of the peaks did not change after irradiation, which indicates that the gain of the SiPM was not affected. This is in agreement with the results reported in [113], where no change of the breakdown voltage was observed for fluences up to $\Phi_{eq} = 5 \cdot 10^{13}$ cm^{-2}. In figure 8.9(b), the optical crosstalk probability is plotted versus the overvoltage. Within the uncertainties, no impact of the irradiation is observed.

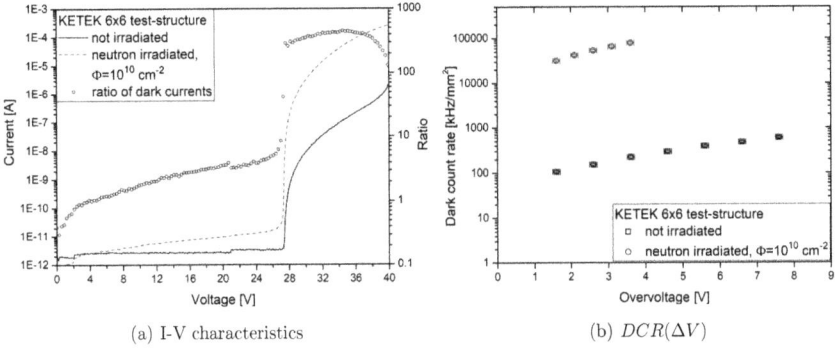

(a) I-V characteristics

(b) $DCR(\Delta V)$

Figure 8.8: Increase of the dark current and the dark count rate of the 6x6 test-structure after the irradiation with neutrons.

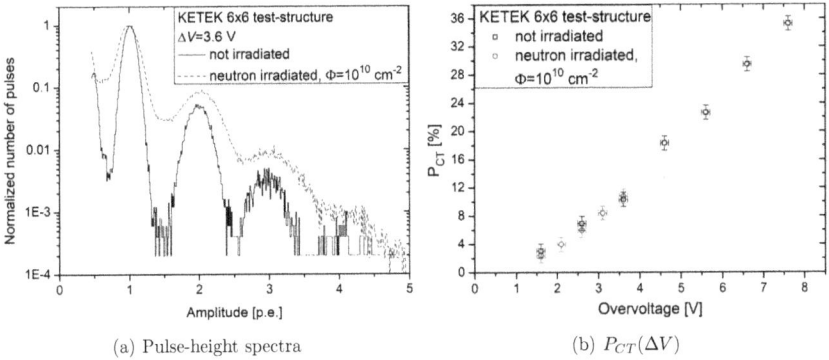

(a) Pulse-height spectra

(b) $P_{CT}(\Delta V)$

Figure 8.9: Impact of the neutron irradiation on the pulse-height spectrum and the optical crosstalk probability of the 6x6 test-structure.

To visualize the defects which were introduced during the irradiation, light emission images were recorded at an applied overvoltage of $\Delta V = 4$ V. For the reference sample, an exposure time of $t_{exp} = 2$ h was chosen, whereas the image of the irradiated sample was recorded with an exposure time of $t_{exp} = 2$ min. The results are shown in figures 8.10(a) and 8.10(b) (note the different intensity scales). The increased number of hotspots after irradiation is evident in these images. Considering the dark count rate of point defects which was discussed in the previous section, the exposure time of 2 min is too small to visualize these kind of crystal damages. For this reason, the observed hotspots in figure 8.10(b) are considered to be cluster defects only.

149

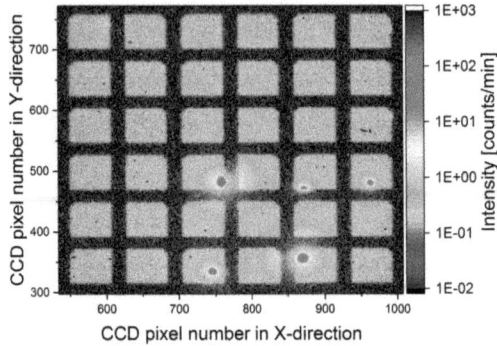

(a) Reference sample, $t_{exp} = 2$ h

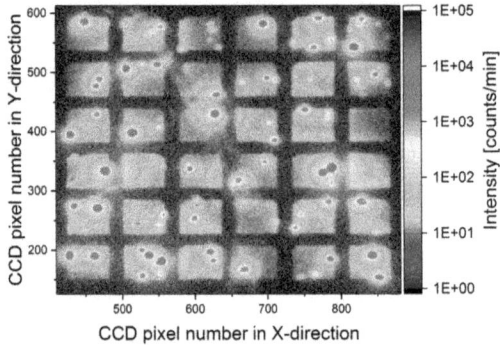

(b) Neutron irradiated sample, $t_{exp} = 2$ min

Figure 8.10: Light emission images of the reference and the neutron irradiated 6x6 test-structures at $\Delta V = 4$ V.

In figures 8.11(a) and 8.11(b), cross-sections of figures 8.10(a) and 8.10(b) through the center of hotspots are shown (note the different intensity scales). It is observed that the amplitudes of the neutron induced defects are about two orders of magnitude larger than ion implantation induced defects (reference sample). From this result it is concluded that the dark count rate generated by neutron induced crystal defects is two orders of magnitude larger that the dark count rate generated by implantation defects. However, one has to consider that the implantation defects are evaluated after an annealing process, whereas the neutron defects are evaluated without a significant annealing. The total CCD pixel amplitude spectra are compared in figure 8.12.

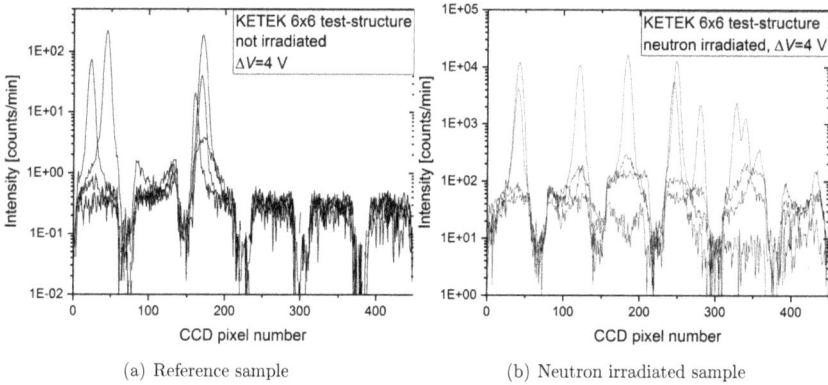

(a) Reference sample   (b) Neutron irradiated sample

Figure 8.11: Comparison of the hotspot amplitudes of the reference and the neutron irradiated 6x6 test-structure in 1D.

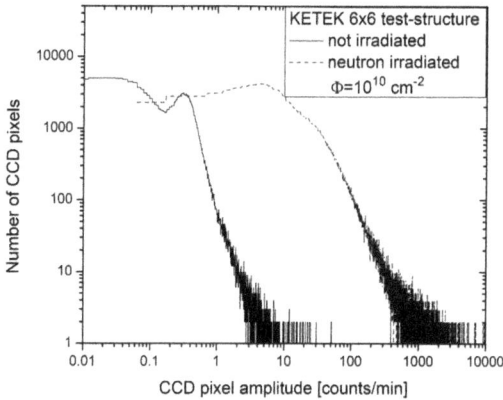

Figure 8.12: CCD pixel amplitude spectra of the reference and the neutron irradiated 6x6 test-structure at $\Delta V = 4$ V.

The temperature dependence of the dark current of the irradiated SiPM was measured in a climate chamber (Vötsch, VCL 7003) from 20 °C to −60 °C. The activation energy was determined by using the conventional method, which is described in section 4.2. In figure 8.13(a), the Arrhenius plots at different overvoltages are shown. In figure 8.13(b), the corresponding activation energy is plotted versus the overvoltage. This activation energy is attributed to the observed hotspots, since the contribution of dark current which is not due to the irradiation damage, can be neglected (see figure 8.8(a)). E_{act} decreases from about 0.28 eV at $\Delta V = 2$ V to about 0.14 eV at $\Delta V = 10$ V. For non-irradiated SiPMs, significantly higher activation energies were observed, as shown in figure 4.5(b). To properly understand this result, it has to be confirmed in additional measurements that the operation of the irradiated SiPM is analogous to the reference SiPM. Due to the high dark current, the voltage drop across the SiPM may change, or self heating may reduce the breakdown voltage. Both effects would lead to a distortion of the presented analysis. At overvoltages $\Delta V < 4$ V, the contribution of these effects is excluded because the gain and the crosstalk probability of the irradiated and the reference SiPM do not differ (see figures 8.9(a) and 8.9(b)).

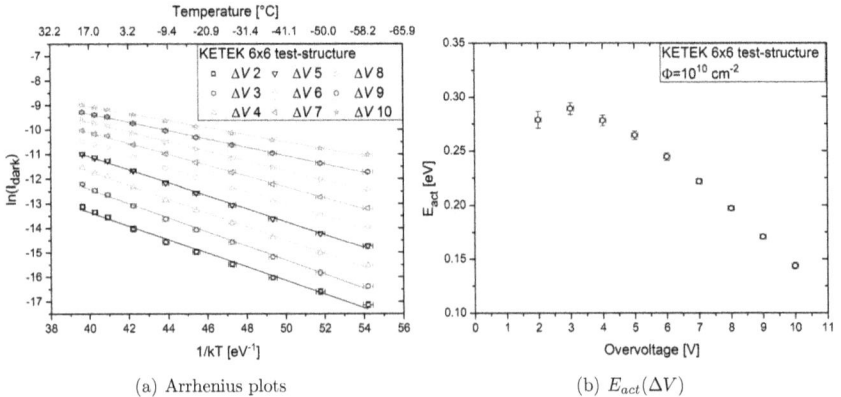

(a) Arrhenius plots

(b) $E_{act}(\Delta V)$

Figure 8.13: Determination of the activation energy of the dark current of the neutron irradiated 6x6 test-structure (conventional method).

8.3 Conclusion

In this chapter, the impact of the irradiation with ^{60}Co γ-rays and the impact of the irradiation with 1 MeV neutrons on the SiPM parameters was discussed. Except for the increased dark count rate, no other parameters were affected by the different radiation types.

Using the low light level camera, a qualitative comparison of the dark count rate of hotspots generated by point defects and cluster defects was performed. The hotspots generated by ion implantation were used as a reference. The dark count rate of point defects generated by the ^{60}Co γ-rays was observed to be more that three orders of magnitude smaller than the dark count rate generated by cluster defects which were created by neutrons.

Using the temperature dependence of the dark current, the activation energy of the neutron induced crystal damages was determined. It was lower than the activation energy determined for non-irradiated SiPMs. Additionally, it showed a strong dependence on the overvoltage, which was also not observed for non-irradiated SiPMs. This result indicates that the charge carrier generation mechanism for neutron damages is not comparable to the mechanism which was observed for the ion implantation induced damages. A detailed analysis of this observation was not performed within the presented research project.

This chapter demonstrates that the experimental methods, which were developed in this work, are well suited for the investigation of the irradiation hardness of SiPMs.

The obtained results already provide useful knowledge and show that lots of exciting and informative experiments are still to come.

Chapter 9

Summary and Outlook

In this thesis, innovative metrological methods for the characterization of the dark count rate of Silicon Photomultipliers were presented. The results obtained by these methods provided the mandatory understanding to achieve a remarkable suppression of the dark count rate of KETEK SiPMs.

For the purpose of a comprehensive characterization of the fundamental SiPM parameters, an algorithm for the signal processing and pulse detection was developed and implemented in LabVIEW and C++. In this thesis, only those parameters were reviewed which were mandatory for the performed analyzes or which were necessary for the lines of argument.

One of the main objectives of the presented research was the identification and evaluation of the physical mechanisms which contribute to the dark count rate of SiPMs. In a conventional approach, the temperature dependence of the dark current was used as an indicator of the physical mechanisms. The activation energy was determined in two temperature regions from the Arrhenius plots of the measured dark currents at fixed overvoltage conditions. The dark current and the dark count rate of KETEK SiPMs are related via the temperature independent gain and the temperature independent optical crosstalk probability. For this reason the determined activation energies are also valid for the dark count rate.

At low overvoltages, this conventional approach indicated the diffusion of charge carriers as the dominating mechanism for dark pulses at temperatures higher than 10 °C. For temperatures lower than −10 °C, the obtained results imply that the measured dark currents are strongly determined by the process of Shockley-Read-Hall-Generation. With increasing overvoltage, a decrease of the activation energy was observed in both temperature regions. This effect was attributed to a non-negligible contribution from electric-field effects. The drawback of the conventional approach is, that the activation energy cannot be extracted at a fixed voltage and overvoltage simultaneously. However, the dark count rate of SiPMs is determined by a superposition of several contributions, some of which depend on the applied voltage and some of which show a functional dependence on the overvoltage. Consequently, the identification of a single mechanism is only possible if all other mechanisms are negligible in a certain temperature range and for certain voltages or overvoltages.

To overcome this restriction, an innovative method for the analysis of the dark current was developed. It is based on the independent measurements of the SiPM photo current and dark current. The photo current is used for the definition of a quantity called "responsivity". The responsivity accounts for the contribution of the overvoltage dependent parameters like the gain, the crosstalk probability, the afterpulsing probability and the avalanche triggering probability. In this way, the physical effects which cause the generation of charge carriers in the multiplication region were separated from the overvoltage specific multiplication of these carriers. With this method it was possible to model the contributions of electric-field effects and the contributions of quasi field-independent mechanisms. Concerning the quasi field-independent part of the dark current I_{ini}, the diffusion of charge carriers was confirmed to be the dominant mechanism at $T \geq 10$ °C. For $T \leq -10$ °C, I_{ini} was found to be solely determined by the Shockley-Read-Hall-Generation. The field-dependent contribution to the dark current was identified with trap-assisted tunneling. For the KETEK PM3350T STD, the relative contribution of trap-assisted tunneling at an overvoltage of 4 V increases from about 30 % at 30 °C to about 60 % at −30 °C. Under the condition of low electric fields, the activation energy determined with the novel method approaches the one determined with the conventional method. This confirms a steady transition between both approaches.

The suppression of the diffusion current was achieved by the introduction of an electric potential to the substrate of the SiPMs. The basic idea was to create an electric field which counteracts the diffusion of majority holes from the p-type substrate into the multiplication region. With this approach, the dark count rate of the KETEK PM3350T STD was reduced from 300 kHz/mm^2 to 100 kHz/mm^2 at room temperature and an overvoltage of $\Delta V = 4$ V. The activation energy of I_{ini} decreased from (1.12 ± 0.03) eV to (0.60 ± 0.02) eV for temperatures between 35 °C and 15 °C. This result confirmed the effective suppression of the diffusion current. At temperatures lower than −5 °C, the dark count rate was not affected. It was assured that the achieved improvement was not accompanied with changes of any other SiPM parameters. The substrate potential was successfully implemented on chip level and is a component of the current commercially available KETEK SiPMs.

The metrological methods which were used up to this point had one common disadvantage. Namely, the lack of information on the location of the generated dark pulses. To overcome this restriction, a novel metrological method was developed within the scope of this work. This method enables a sub-micro-cell, 2D spatially resolved measurement of the dark count rate within the plane of the active area. The basic principle of this method is to detect and map the light which is emitted during the avalanche breakdown of micro-cells with a low-light-level CCD camera. Due to the existing functional dependency of the dark count rate and the emitted light, which was demonstrated in this thesis, the light intensity map is equivalent to a map of the dark count rate. The application of this method revealed sub-micro-cell regions with a strongly increased charge carrier generation rate (hotspots) for SiPMs produced by KETEK, SensL and Hamamatsu.

For KETEK SiPMs, an extended characterization of hotspots was performed in the course of the presented research project. It was found that for the PM3350T STD-SP, 90 % of the most intense hotspots are responsible for approximately 55 % of the total dark count rate. Using the temperature dependence of the spatially resolved dark count rate, the activation energies were determined in the areas attributed to hotspots and in hotspot-free areas. The physical mechanism responsible for hotspots was found to be a combination of the Shockley-Read-Hall-Generation and trap-assisted tunneling.

The hotspots were identified as crystal defects generated during the implantation of ionized phosphorus which forms the buried n-layer. It was demonstrated that an effective suppression of the crystal defect density in the active region of micro-cells can be achieved by two distinct approaches. The first approach is to reduce the phosphorus peak concentration. Here, a reduction of the dark count rate from 100 kHz/mm^2 to 70 kHz/mm^2 at room temperature and an overvoltage of $\Delta V = 4$ V was achieved by reducing the peak concentration by a factor of two (PM4450T LD-SP). The second approach is to shift the position of the phosphorus peak concentration to larger depths within the silicon. This can either be realized by increasing the implantation energy or by using a thinner scatter oxide. By increasing the implantation energy from 3.5 MeV (PM3350T STD-SP) to 4.5 MeV (PM3350T HE-SP), a suppression of the total dark count rate down to 40 kHz/mm^2 at room temperature and an overvoltage of $\Delta V = 4$ V was achieved. By compensating the higher implantation energy with a thicker scatter oxide (PM3350T HE-S-SP), the dark count rate of the PM3350T STD-SP was reproduced. This result confirms that the explicit dependence of the crystal defect density on the implantation energy can be neglected. Again, the achieved improvement did not change other SiPM parameters, except for the breakdown voltage due to the shift of the phosphorus distribution.

The analysis of the activation energy in the hotspot-free areas revealed a remaining diffusion current. It is assumed that this current is due to minority holes diffusing from the non-depleted n-layer into the active region. The relative contribution of the remaining diffusion current is increased for the PM3350T HE-SP with respect to the PM3350T STD-SP. This is in agreement with the reduced crystal defect density in the active region of the PM3350T HE-SP.

For certain micro-cells, an enhanced dark count rate was observed in the corner in which the cathode contact is integrated. It is assumed that the enhanced electric field in this region is responsible for this effect. Preliminary results indicate that the affected micro-cell corners are responsible for about 10 % of the total dark count rate. Samples for which the cathode contact was moved from the corner to a central position along the edge of the micro-cell did not show such a contribution. During the presented research project, it was not possible to investigate this effect in more detail to better understand its underlying physics. Consequently, the obtained preliminary results were not reviewed in this thesis. However, it is an interesting and important topic for future research projects.

During the application of SiPMs in radiation hard environments, the dark count rate is the SiPM parameter which degrades the fastest and determines the lifetime of this photon counter. In this work, the radiation damages in KETEK SiPMs were investigated. Therefor, the samples were irradiated with different doses of ^{60}Co γ-rays and 1 MeV neutrons with a fluence of $\Phi = 10^{10}$ cm^{-2}. At an overvoltage of 4 V, the PM1150T STD-SP showed a linear increase of the dark count rate from 100 kHz/mm^2 to 700 kHz/mm^2 with the accumulated ^{60}Co γ-ray dose extending from 40 Gy to 240 Gy. In a preliminary technology study, the PM3350T HE-SP showed a higher radiation tolerance to 120 Gy of ^{60}Co γ-rays than the PM3350T STD-SP. At an overvoltage of 4 V, the difference of the absolute increase of the dark count rate differed by approximately 80 kHz/mm^2 in favor of the PM3350T HE-SP. A possible explanation of this result may be related with the different distributions of the avalanche triggering probabilities of both SiPM types. Two approaches for the analysis of these distributions were proposed for future experiments.

After the neutron irradiation, the dark count rate of the investigated sample increased by a factor of about 400. Preliminary results showed a lower activation energy for the dark count rate of the neutron irradiated SiPM with respect to the non-irradiated devices. This indicates, that the charge carrier generation mechanism of neutron-induced defects differs from the one of ion implantation defects. For a better understanding of this observation, additional research is necessary.

Using the low-light-level camera, the dark count rate of point defects due to ^{60}Co γ-rays and the dark count rate of cluster defects due to 1 MeV neutrons was evaluated respectively. The obtained results reveal, that the dark count rate of a cluster defect is more than three orders of magnitude higher that the dark count rate of a point defect. Neither after the γ-ray irradiation nor after the neutron irradiation a change of the breakdown voltage, the optical crosstalk probability or the relative gain of the SiPMs was observed.

One of the current goals in SiPM development is to reduce the dark count rate to the benchmark of < 10 kHz/mm^2 at the condition of a saturated photon detection efficiency. To achieve this goal, the crystal defects in the active region have to be further reduced. Additionally, the distribution of the electric field has to be optimized to obtain the maximum avalanche triggering probability with a simultaneously low field-assisted generation of charge carriers in the absence of photons.

The promising potential of the Silicon Photomultiplier has not been fully exhausted yet.

Bibliography

[1] D. Renker, E. Lorenz, "Advances in solid state photon detectors", Journal of Instrumentation 4 (04) (2009) P04004–P04004. `doi:10.1088/1748-0221/4/04/P04004`. 2, 10, 13, 21, 54

[2] N. Dinu, "8 - Silicon photomultipliers (SiPM) ", in: B. Nabet (Ed.), Photodetectors, Woodhead Publishing, 2016, pp. 255 – 294. `doi:10.1016/B978-1-78242-445-1.00008-7`. 2

[3] D. Durini, U. Paschen, A. Schwinger, A. Spickermann, "11 - Silicon based single-photon avalanche diode (SPAD) technology for low-light and high-speed applications", in: B. Nabet (Ed.), Photodetectors, Woodhead Publishing, 2016, pp. 345 – 371. `doi:10.1016/B978-1-78242-445-1.00011-7`. 2

[4] J. L. Wiza, "Microchannel plate detectors", Nuclear Instruments and Methods 162 (1) (1979) 587 – 601. `doi:10.1016/0029-554X(79)90734-1`. 2

[5] G. Gasanov, V. Golovin, Z. Sadygov, N. Yusipov, Russian patent #1702831,1989. 2

[6] V. Golovin, Z. Sadygov, M. Tarasov, N. Yusipov , Russian patent #1644708,1989. 2

[7] KETEK, "Recent Enhancements of the KETEK SiPM Device Performance with regard to Timing, Cross Talk and CMOS Compatibility", in: Talk given at PhotoDet, International Conference on New Photo-detectors, Troitsk, Russia, 2015. 2

[8] M. Teshima, B. Dolgoshein, R. Mirzoyan, Nincovic, E. Popova, "SiPM Development for Astroparticle Physics Applications" `arXiv:0709.1808`. 2

[9] M. Shayduk, R. Mirzoyan, M. Kurz, M. Knötig, J. Bolmont, H. Dickinson, E. Lorenz, J.-P. Tavernet, J. Hose, M. Teshima, P. Vincent, "Light sensors selection for the Cherenkov Telescope Array: PMT and SiPM", Nuclear Instruments and Methods in Physics Research Section A: Accelerators, Spectrometers, Detectors and Associated Equipment 695 (2012) 109–112. `doi:10.1016/j.nima.2011.12.010`. 2

[10] N. Otte, B. Dolgoshein, J. Hose, S. Klemin, E. Lorenz, R. Mirzoyan, E. Popova, M. Teshima, "The Potential of SiPM as Photon Detector in Astroparticle Physics Experiments like MAGIC and EUSO", Nuclear Physics B - Proceedings Supplements 150 (2006) 144–149. doi:10.1016/j.nuclphysbps.2004.10.084. 2, 35

[11] B. Majorovits for the GERDA collaboration, "The search for $0\eta\beta\beta$ decay with the GERDA experiment: status and prospects"arXiv:1506.00415v1. 2

[12] I. Ostrovskiy, F. Retiere, D. Auty, J. Dalmasson, T. Didberidze, R. DeVoe, G. Gratta, L. Huth, L. James, L. Lupin-Jimenez, N. Ohmart, A. Piepke, "Characterization of Silicon Photomultipliers for nEXO", IEEE Transactions on Nuclear Science 62 (4) (2015) 1825–1836. doi:10.1109/TNS.2015.2453932. 2

[13] T. Igarashi, M. Tanaka, T. Washimi, K. Yorita, "Performance of VUV-sensitive MPPC for liquid argon scintillation light", Nuclear Instruments and Methods in Physics Research Section A: Accelerators, Spectrometers, Detectors and Associated Equipment 833 (Supplement C) (2016) 239 – 244. doi:10.1016/j.nima.2016.07.008. 2

[14] S. Ogawa, "Liquid xenon calorimeter for MEG II experiment with VUV-sensitive MPPCs", Nuclear Instruments and Methods in Physics Research Section A: Accelerators, Spectrometers, Detectors and Associated Equipment 845 (Supplement C) (2017) 528 – 532, proceedings of the Vienna Conference on Instrumentation 2016. doi:10.1016/j.nima.2016.06.085. 2

[15] J. Janicskó Csáthy, H. Aghaei Khozani, a. Caldwell, X. Liu, B. Majorovits, "Development of an anti-Compton veto for HPGe detectors operated in liquid argon using silicon photo-multipliers", Nuclear Instruments and Methods in Physics Research Section A: Accelerators, Spectrometers, Detectors and Associated Equipment 654 (1) (2011) 225–232. doi:10.1016/j.nima.2011.05.070. 3

[16] V. Kushpil, V. Mikhaylov, A. Kugler, S. Kushpil, V. P. Ladygin, O. Svoboda, P. Tlustý, "Radiation hardness of semiconductor avalanche detectors for calorimeters in future HEP experiments", Journal of Physics: Conference Series 675 (1) (2016) 012039. doi:10.1088/1742-6596/675/1/012039. 3

[17] N. Strobbe, "The upgrade of the CMS hadron calorimeter with silicon photomultipliers", Journal of Instrumentation 12 (01) (2017) C01080.
URL http://stacks.iop.org/1748-0221/12/i=01/a=C01080 3

[18] B. Lutz, the CMS collaboration), "Upgrade of the CMS Hadron Outer Calorimeter with SiPM sensors", Journal of Physics: Conference Series 404 (1) (2012) 012018. `doi:10.1088/1742-6596/404/1/012018`. 3, 141

[19] E. Garutti, "Silicon photomultipliers for high energy physics detectors", Journal of Instrumentation 6 (10) (2011) C10003–C10003. `arXiv:1108.3166, doi:10.1088/1748-0221/6/10/C10003`. 3, 141

[20] A. Lobanov, the CMS Collaboration, "The CMS Outer HCAL SiPM Upgrade", Journal of Physics: Conference Series 587 (1) (2015) 012005. `doi:088/1742-6596/587/1/012005`. 3

[21] M. C. Vignali, V. Chmill, E. Garutti, R. Klanner, M. Nitschke, J. Schwandt, S. Sonder, "Neutron induced radiation damage of KETEK SiPMs", in: 2016 IEEE Nuclear Science Symposium, Medical Imaging Conference and Room-Temperature Semiconductor Detector Workshop (NSS/MIC/RTSD), 2016, pp. 1–5. `doi:10.1109/NSSMIC.2016.8069733`. 3

[22] A. Gola et al., "Characterization of FBK RGB-UHD SiPMs Ulter High Density", in: Talk given at PhotoDet, International Conference on New Photo-detectors, Troitsk, Russia, 2015. 3

[23] Y. Musienko, "Status and perspectives of solid state photon detectors", in: Talk given at RICH, International Workshop on Ring Imaging Cherenkov Detectors, Bled, Slovenia, 2016. 3

[24] A. Ferri, F. Acerbi, A. Gola, C. Piemonte, G. Paternoster, N. Zorzi, "Performance of a 64-channel, $3.2x3.2cm^2$ SiPM tile for TOF-PET application", Nuclear Instruments and Methods in Physics Research Section A: Accelerators, Spectrometers, Detectors and Associated Equipment 824 (Supplement C) (2016) 196 – 197, frontier Detectors for Frontier Physics: Proceedings of the 13th Pisa Meeting on Advanced Detectors. `doi:10.1016/j.nima.2015.11.084`. 4

[25] S. Gundacker, E. Auffray, B. Frisch, H. Hillemanns, P. Jarron, T. Meyer, K. Pauwels, P. Lecoq, "A Systematic Study to Optimize SiPM Photo-Detectors for Highest Time Resolution in PET", IEEE Transactions on Nuclear Science 59 (5) (2012) 1798–1804. `doi:10.1109/TNS.2012.2202918`. 4

[26] M. F. Santangelo, D. Sanfilippo, G. Fallica, A. C. Busacca, R. Pagano, E. L. Sciuto, S. Lombardo, S. Libertino, "SiPM as novel optical biosensor transduction and applications", in: 2014 Fotonica AEIT Italian Conference on Photonics Technologies, 2014, pp. 1–4. `doi:10.1109/Fotonica.2014.6843944`. 4

[27] M. Goesch, A. Serov, A. Rochas, H. Blom, T. Anhut, P. Besse, R. Popovic, T. Lasser, R. Rigler, "Parallel single molecule detection with a fully integrated single-photon 2x2 CMOS detector array", Journal of Biomedical Optics 9 (2004) 913–921. `doi:10.1117/1.1781668`. 4

[28] H. C. G. Alexandra V. Agronskaia, L. Tertoolen, "Fast fluorescence lifetime imaging of calcium in living cells", Journal of Biomedical Optics 9 (2004) 1230–1237. `doi:10.1117/1.1806472`. 4

[29] R. Daniel, R. Almog, A. Ron, S. Belkin, Y. S. Diamand, "Modeling and measurement of a whole-cell bioluminescent biosensor based on a single photon avalanche diode", Biosensors and Bioelectronics 24 (4) (2008) 882 – 887. `doi:10.1016/j.bios.2008.07.026`. 4

[30] E. Grigoriev, A. Akindinov, M. Breitenmoser, S. Buono, E. Charbon, C. Niclass, I. Desforges, R. Rocca, "Silicon photomultipliers and their bio-medical applications", Nuclear Instruments and Methods in Physics Research Section A: Accelerators, Spectrometers, Detectors and Associated Equipment 571 (1) (2007) 130 – 133, proceedings of the 1st International Conference on Molecular Imaging Technology. `doi:10.1016/j.nima.2006.10.046`. 4

[31] H. Li, N. Lopes, S. Moser, G. Sayler, S. Ripp, "Silicon photomultiplier (SPM) detection of low-level bioluminescence for the development of deployable whole-cell biosensors: Possibilities and limitations", Biosensors and Bioelectronics 33 (1) (2012) 299 – 303. `doi:10.1016/j.bios.2012.01.008`. 4

[32] F. Corsi, A. Dragone, C. Marzocca, A. Del Guerra, P. Delizia, N. Dinu, C. Piemonte, M. Boscardin, G. F. Dalla Betta, "Modelling a silicon photomultiplier (SiPM) as a signal source for optimum front-end design", Nuclear Instruments and Methods in Physics Research, Section A: Accelerators, Spectrometers, Detectors and Associated Equipment 572 (1 SPEC. ISS.) (2007) 416–418. `doi:10.1016/j.nima.2006.10.219`. 10

[33] S. Seifert, H. T. van Dam, J. Huizenga, R. Vinke, P. Dendooven, H. Lohner, D. R. Schaart, "Simulation of Silicon Photomultiplier Signals", IEEE Transactions on Nuclear Science 56 (6) (2009) 3726–3733. doi:10.1109/TNS.2009.2030728. 10, 12

[34] F. Corsi, M. Foresta, C. Marzocca, G. Matarrese, A. D. Guerra, "Current-mode front-end electronics for silicon photo-multiplier detectors", in: 2007 2nd International Workshop on Advances in Sensors and Interface, 2007, pp. 1–6. doi: 10.1109/IWASI.2007.4420025. 12

[35] "Linear Technology, SPICE simulator LTSpice", online available at: http://www. linear.com/designtools/software/ (accessed December 08, 2017). 12

[36] P. Buzhan, B. Dolgoshein, A. Ilyin, V. Kantserov, V. Kaplin, A. Karakash, A. Pleshko, E. Popova, S. Smirnov, Y. Volkov, L. Filatov, S. Klemin, F. Kayumov, "The Advanced Study of Silicon Photomultiplier", ICFA Instrumentation Bulletin 23. doi:10.1142/ 9789812776464_0101. 13

[37] C. L. Anderson, C. R. Crowell, "Threshold Energies for electron-hole pair production by impact ionization in semiconductor", Phys. Rev. B 5 (100) (1972) 2267–2272. doi: 10.1103/PhysRevB.5.2267. 14

[38] P. A. Wolff, "Theory of Electron Multiplication in Silicon and Germanium", Phys. Rev. 95 (1954) 1415–1420. doi:10.1103/PhysRev.95.1415. 15

[39] W. Shockley, "Problems related to p-n junctions in silicon", Solid-State Electronics 2 (1) (1961) 35 – 67. doi:10.1016/0038-1101(61)90054-5. 15

[40] W. J. Kindt, "Geiger Mode Avalanche Photodiode Arrays for Spatially Resolved Single Photon Counting", Delft University Press, 1999. 16, 19, 29, 30, 36

[41] R. Van Overstraeten, H. De Man, "Measurement of the ionization rates in diffused silicon p-n junctions", Solid State Electronics 13 (5) (1970) 583–608. doi:10.1016/ 0038-1101(70)90139-5. 16

[42] R. J. McIntyre, "Multiplication noise in uniform avalanche diodes", IEEE Transactions on Electron Devices ED-13 (1) (1966) 164–168. doi:10.1109/T-ED.1966.15651. 17

[43] P. Finocchiaro, A. Pappalardo, L. Cosentino, M. Belluso, S. Billotta, G. Bonanno, B. Carbone, G. Condorelli, S. D. Mauro, G. Fallica, M. Mazzillo, A. Piazza, D. Sanfilippo, G. Valvo, "Characterization of a Novel 100-Channel Silicon Photomultiplier

- Part I: Noise", IEEE Transactions on Electron Devices 55 (10) (2008) 2757–2764. doi:10.1109/TED.2008.2003996. 18

[44] G. Condorelli, D. Sanfilippo, G. Valvo, M. Mazzillo, D. Bongiovanni, a. Piana, B. Carbone, G. Fallica, "Extensive electrical model of large area silicon photomultipliers", Nuclear Instruments and Methods in Physics Research Section A: Accelerators, Spectrometers, Detectors and Associated Equipment 654 (1) (2011) 127–134. doi:10.1016/j.nima.2011.07.034. 19

[45] R. J. Mclntyre, "A new look at impact Ionization-part I: A theory of gain, noise, breakdown probability, and frequency response", IEEE Transactions on Electron Devices 46 (8) (1999) 1623–1631. doi:10.1109/16.777150. 20

[46] W. G. Oldham, R. R. Samuelson, P. Antognetti, "Triggering phenomena in avalanche diodes", IEEE Transactions on Electron Devices 19 (9) (1972) 1056–1060. doi:10.1109/T-ED.1972.17544. 20, 21

[47] R. H. Haitz, "Mechanisms contributing to the noise pulse rate of avalanche diodes", Journal of Applied Physics 36 (10) (1965) 3123–3131. doi:10.1063/1.1702936. 21

[48] R. Pagano, D. Corso, S. Lombardo, G. Valvo, D. N. Sanfilippo, G. Fallica, S. Libertino, "Dark Current in Silicon Photomultiplier Pixels: Data and Model", IEEE Transactions on Electron Devices 59 (9) (2012) 2410–2416. doi:10.1109/TED.2012.2205689. 21, 63

[49] G. Hurkx, D. Klaassen, M. Knuvers, "A new recombination model for device simulation including tunneling", IEEE Transactions on Electron Devices 39 (2) (1992) 331–338. doi:10.1109/16.121690. 21, 26

[50] S. M. Sze, K. K. Ng, "Physics of Semiconductor Devices", 3rd Edition, John Wiley & Sons, Inc., 2007. arXiv:9809069, doi:10.1080/01422419908228843. 23, 24

[51] G. E. Stillman, C. M. Wolfe, J. O. Dimmock, "Hall coefficient factor for polar mode scattering in n-type GaAs", Journal of Physics and Chemistry of Solids 31 (6) (1970) 1199–1204. doi:10.1016/0022-3697(70)90122-8. 24

[52] D. K. Schroder, "Semiconductor Material and Device Characterization: Third Edition", 2005. doi:10.1002/0471749095. 24, 68

166

[53] V. Alex, S. Finkbeiner, J. Weber, "Temperature dependence of the indirect energy gap in crystalline silicon", Journal of Applied Physics 79 (9) (1996) 6943–6946. doi: 10.1063/1.362447. 24

[54] D. K. Schroder, "Carrier lifetimes in silicon", IEEE Transactions on Electron Devices 44 (1) (1997) 160–170. doi:10.1109/16.554806. 25

[55] S. Ganichev, E. Ziemann, W. Prettl, I. Yassievich, a. Istratov, E. Weber, "Distinction between the Poole-Frenkel and tunneling models of electric-field-stimulated carrier emission from deep levels in semiconductors", Physical Review B 61 (15) (2000) 10361–10365. doi:10.1103/PhysRevB.61.10361. 28, 29

[56] G. Hurkx, H. de Graaff, W. Kloosterman, M. Knuvers, "A new analytical diode model including tunneling and avalanche breakdown", IEEE Transactions on Electron Devices 39 (9) (1992) 2090–2098. doi:10.1109/16.155882. 29

[57] R. Mirzoyan, R. Kosyra, H. G. Moser, "Light emission in Si avalanches", Nuclear Instruments and Methods in Physics Research, Section A: Accelerators, Spectrometers, Detectors and Associated Equipment 610 (1) (2009) 98–100. doi:10.1016/j.nima. 2009.05.081. 30, 31

[58] R. Newman, "Visible Light from a Silicon $p - n$ Junction", Phys. Rev. 100 (1955) 700–703. doi:10.1103/PhysRev.100.700. 31

[59] N. Akil, S. E. Kerns, D. V. Kerns, A. Hoffmann, J. P. Charles, "Photon generation by silicon diodes in avalanche breakdown", Applied Physics Letters 73 (7) (1998) 871–872. doi:10.1063/1.121971. 31, 32, 33

[60] D. K. Gautam, W. S. Khokle, K. B. Garg, "Photon emission from reverse-biased silicon P-N junctions", Solid State Electronics 31 (2) (1988) 219–222. doi:10.1016/0038-1101(88)90130-X. 31

[61] P. Wolff, "Theory of optical radiation from breakdown avalanches in germanium", Journal of Physics and Chemistry of Solids 16 (3-4) (1960) 184–190. doi:10.1016/0022-3697(60)90148-7. 31

[62] A. Toriumi, M. Yoshimi, M. Iwase, Y. Akiyama, K. Taniguchi, "A Study of Photon Emission from n-Channel MOSFET's", IEEE Transactions on Electron Devices 34 (7) (1987) 1501–1508. doi:10.1109/T-ED.1987.23112. 32

[63] A. L. Lacaita, F. Zappa, S. Bigliardi, M. Manfredi, "On the Bremsstrahlung Origin of Hot-Carrier-Induced Photons in Silicon Devices", IEEE Transactions on Electron Devices 40 (3) (1993) 577–582. doi:10.1109/16.199363. 33, 34, 103

[64] J. Bude, N. Sano, A. Yoshii, "Hot-carrier luminescence in Si", Physical Review B 45 (1992) 5848–5856. doi:10.1103/PhysRevB.45.5848. 33

[65] F. Retière, K. Boone, "Delayed avalanches in Multi-Pixel Photon Counters", in: 2012 IEEE Nuclear Science Symposium and Medical Imaging Conference Record (NSS/MIC), 2012, pp. 1585–1588. doi:10.1109/NSSMIC.2012.6551378. 35

[66] R. Mirzoyan, B. Dolgoshein, P. Holl, S. Klemin, C. Merck, H.-G. Moser, A. N. Otte, J. Ninković, E. Popova, R. Richter, M. Teshima, "SiPM and ADD as advanced detectors for astro-particle physics", Nuclear Instruments and Methods in Physics Research Section A: Accelerators, Spectrometers, Detectors and Associated Equipment 572 (1) (2007) 493–494. doi:10.1016/j.nima.2006.10.151. 35

[67] P. Buzhan, B. Dolgoshein, A. Ilyin, V. Kaplin, S. Klemin, R. Mirzoyan, E. Popova, M. Teshima, "The cross-talk problem in SiPMs and their use as light sensors for imaging atmospheric Cherenkov telescopes", Nuclear Instruments and Methods in Physics Research Section A: Accelerators, Spectrometers, Detectors and Associated Equipment 610 (1) (2009) 131–134. doi:10.1016/j.nima.2009.05.150. 35

[68] F. Nagy, M. Mazzillo, L. Renna, G. Valvo, D. Sanfilippo, B. Carbone, A. Piana, G. Fallica, J. Molnár, "Afterpulse and delayed crosstalk analysis on a STMicroelectronics silicon photomultiplier", Nuclear Instruments and Methods in Physics Research Section A: Accelerators, Spectrometers, Detectors and Associated Equipment 759 (Supplement C) (2014) 44 – 49. doi:10.1016/j.nima.2014.04.045.
URL http://www.sciencedirect.com/science/article/pii/S0168900214004501
35

[69] C. Dietzinger, P. Iskra, T. Ganka, T. Eggert, L. Höllt, A. Pahlke, N. Miyakawa, M. Fraczek, J. Knobloch, F. Wiest, W. Hansch, R. Fojt, "Reduction of optical crosstalk in silicon photomultipliers" (2012). doi:10.1117/12.930473. 35

[70] V. Saveliev, "Silicon Photomultiplier - New Era of Photon Detection", in: K. Y. Kim (Ed.), Advances in Optical and Photonic Devices, InTech, Rijeka, 2010, Ch. 14. doi:10.5772/7150. 36

[71] H. T. van Dam, S. Seifert, D. R. Schaart, "The statistical distribution of the number of counted scintillation photons in digital silicon photomultipliers: model and validation", Physics in Medicine and Biology 57 (15) (2012) 4885–4903. doi:10.1088/0031-9155/57/15/4885. 36

[72] B. Seitz, A. G. Stewart, K. O'Neill, L. Wall, C. Jackson, "Performance evaluation of novel SiPM for medical imaging applications", in: 2013 IEEE Nuclear Science Symposium and Medical Imaging Conference (2013 NSS/MIC), 2013, pp. 1–4. doi:10.1109/NSSMIC.2013.6829685. 36

[73] D. Henseler, R. Grazioso, N. Zhang, M. Schmand, "SiPM performance in PET applications: An experimental and theoretical analysis", in: 2009 IEEE Nuclear Science Symposium Conference Record (NSS/MIC), 2009, pp. 1941–1948. doi:10.1109/NSSMIC.2009.5402157. 36

[74] S. Vinogradov, T. Vinogradova, V. Shubin, D. Shushakov, C. Sitarsky, "Efficiency of Solid State Photomultipliers in Photon Number Resolution", IEEE Transactions on Nuclear Science 58 (1) (2011) 9–16. doi:10.1109/TNS.2010.2096474. 36

[75] S. Cova, A. Lacaita, G. Ripamonti, "Trapping Phenomena in Avalanche Photodiodes on Nanosecond Scale", IEEE Electron Device Letters 12 (12) (1991) 685–687. doi:10.1109/55.116955. 36

[76] H. Oide, H. Otono, H. Hano, T. Suehiro, S. Yamashita, T. Yoshioka, "Study of afterpulsing of MPPC with waveform analysis", in: International Workshop on New Photon-Detectors, Kobe, Japan, Vol. PD07, 2007. 36

[77] J. Stein, F. Scheuer, W. Gast, A. Georgiev, "X-ray detectors with digitized preamplifiers", Nuclear Instruments and Methods in Physics Research Section B: Beam Interactions with Materials and Atoms 113 (1-4) (1996) 141–145. doi:10.1016/0168-583X(95)01417-9. 40

[78] N. Dinu, R. Battiston, M. Boscardin, G. Collazuol, F. Corsi, G. F. Dalla Betta, A. Del Guerra, G. Llosá, M. Ionica, G. Levi, S. Marcatili, C. Marzocca, C. Piemonte, G. Pignatel, A. Pozza, L. Quadrani, C. Sbarra, N. Zorzi, "Development of the first prototypes of Silicon PhotoMultiplier (SiPM) at ITC-irst", Nuclear Instruments and Methods in Physics Research, Section A: Accelerators, Spectrometers, Detectors and Associated Equipment 572 (1 SPEC. ISS.) (2007) 422–426. doi:10.1016/j.nima.2006.10.305. 41

[79] V. Chmill, E. Garutti, R. Klanner, M. Nitschke, J. Schwandt, "Study of the breakdown voltage of SiPMs", Nuclear Instruments and Methods in Physics Research, Section A: Accelerators, Spectrometers, Detectors and Associated Equipment 845 (2017) 56–59. arXiv:1605.01692, doi:10.1016/j.nima.2016.04.047. 43, 44

[80] E. Garutti, M. Ramilli, C. Xu, L. Hellweg, "Characterization and X-Ray damage of Silicon Photomultipliers", in: Technology and Instrumentation in Particle Physics, Amsterdam, Netherlands, Vol. PoS(TIPP2014)070, 2014, pp. 1–6. 43

[81] P. Eckert, H.-C. Schultz-Coulon, W. Shen, R. Stamen, A. Tadday, "Characterisation studies of silicon photomultipliers", Nuclear Instruments and Methods in Physics Research Section A: Accelerators, Spectrometers, Detectors and Associated Equipment 620 (2) (2010) 217 – 226. doi:10.1016/j.nima.2010.03.169. 47, 54

[82] A. N. Otte, D. Garcia, T. Nguyen, D. Purushotham, "Characterization of Three High Efficiency and Blue Sensitive Silicon Photomultipliers", Nuclear instruments & methods in physics research. Section A, Accelerators, spectrometers, detectors and associated equipment 14 (8) (2016) 1–21. arXiv:1606.05186, doi:10.1016/j.nima.2016.09.053. 47, 48, 54

[83] S. Vinogradov, "Precise Metrology of SiPM: Measurement and Reconstruction of Time Distributions of Single Photon Detections and Correlated Events", in: Talk given at IEEE NSS/MIC/RTSD, Strasbourg, 2016. 50, 51

[84] Y. Du, F. Retière, "After-pulsing and cross-talk in multi-pixel photon counters", Nuclear Instruments and Methods in Physics Research Section A: Accelerators, Spectrometers, Detectors and Associated Equipment 596 (3) (2008) 396–401. doi:10.1016/j.nima.2008.08.130. 53

[85] K. O'Neill, C. Jackson, "SensL B-Series and C-Series silicon photomultipliers for time-of-flight positron emission tomography", Nuclear Instruments and Methods in Physics Research, Section A: Accelerators, Spectrometers, Detectors and Associated Equipment 787 (2015) 169–172. doi:10.1016/j.nima.2014.11.087. 54

[86] L. Futlik, E. Levin, S. Vinogradov, V. Shubin, D. Shushakov, K. Sitarskii, E. Shelegeda, "Methodical problems of crosstalk probability measurements in solid-state photomultipliers", Bulletin of the Lebedev Physics Institute 38 (10) (2011) 302–310. doi:10.3103/S1068335611100058. 55

[87] A. Vacheret, G. Barker, M. Dziewiecki, P. Guzowski, M. Haigh, B. Hartfiel, A. Iz-maylov, W. Johnston, M. Khabibullin, A. Khotjantsev, Y. Kudenko, R. Kurjata, T. Kutter, T. Lindner, P. Masliah, J. Marzec, O. Mineev, Y. Musienko, S. Oser, F. Retiere, R. Salih, A. Shaikhiev, L. Thompson, M. Ward, R. Wilson, N. Yershov, K. Zaremba, M. Ziembicki, "Characterization and simulation of the response of Multi-Pixel Photon Counters to low light levels", Nuclear Instruments and Methods in Physics Research Section A: Accelerators, Spectrometers, Detectors and Associated Equipment 656 (1) (2011) 69–83. doi:10.1016/j.nima.2011.07.022. 55

[88] G. Collazuol, M. Bisogni, S. Marcatili, C. Piemonte, a. Del Guerra, "Studies of silicon photomultipliers at cryogenic temperatures", Nuclear Instruments and Methods in Physics Research Section A: Accelerators, Spectrometers, Detectors and Associated Equipment 628 (1) (2011) 389–392. doi:10.1016/j.nima.2010.07.008. 55

[89] Optometrics, Optometrics Tunable Light Sources TLS6, online available at: https://www.dynasil.com/product-category/tunable-light-sources/ (accessed December 08, 2017). 61

[90] H. A. Weakliem, D. Redfield, "Temperature dependence of the optical properties of silicon", Journal of Applied Physics 50 (3) (1979) 1491–1493. doi:10.1063/1.326135. 61

[91] A. Poyai, E. Simoen, C. Claeys, A. Czerwinski, E. Gaubas, "Improved extraction of the activation energy of the leakage current in silicon p-n junction diodes", Applied Physics Letters 78 (14) (2001) 1997–1999. doi:10.1063/1.1359487. 63

[92] N. Dinu, C. Bazin, V. Chaumat, C. Cheikali, A. Para, V. Puill, C. Sylvia, J. F. Vagnucci, "Temperature and bias voltage dependence of the MPPC detectors", IEEE Nuclear Science Symposium Conference Record (2) (2010) 215–219. doi:10.1109/NSSMIC.2010.5873750. 63

[93] E. Auffray, F. B. M. B. Hadj, D. Cortinovis, K. Doroud, E. Garutti, P. Lecoq, Z. Liu, R. Martinez, M. Paganoni, M. Pizzichemi, A. Silenzi, C. Xu, M. Zvolský, "Characterization studies of silicon photomultipliers and crystals matrices for a novel time of flight PET detector", Journal of Instrumentation 10 (06) (2015) P06009. arXiv:1501.04233v1. 63

[94] S. Vinogradov, "Analytical models of probability distribution and excess noise factor of solid state photomultiplier signals with crosstalk", Nuclear Instruments and Methods

in Physics Research, Section A: Accelerators, Spectrometers, Detectors and Associated Equipment 695 (2012) 247–251. `arXiv:1109.2014`, `doi:10.1016/j.nima.2011.11.086`. 64

[95] R. Pagano, S. Libertino, D. Corso, S. Lombardo, G. Valvo, D. Sanfilippo, M. Mazzillo, A. Piana, G. Fallica, "Silicon Photomultiplier : Technology Improvement and Performance", International Journal on Advances in Systems and Measurements 6 (2013) 124–136. 79

[96] Andor Clara, Datasheet, online available at: `http://www.andor.com/pdfs/specifications/Andor_Clara_Series_Specifications.pdf` (accessed December 08, 2017). 93, 94, 112

[97] T. Frach, G. Prescher, C. Degenhardt, R. de Gruyter, A. Schmitz, R. Ballizany, "The digital silicon photomultiplier - Principle of operation and intrinsic detector performance", in: IEEE Nuclear Science Symposium Conference Record (NSS/MIC), 2009, pp. 1959–1965. `doi:10.1109/NSSMIC.2009.5402143`. 95

[98] E. Popova et al., "Simulation and measurements of Geiger discharge transverse size in a SiPM cell", in: Talk given at IEEE NSS/MIC/RTSD, Seoul, Korea (South), 2013. 100

[99] SRIM - The Stopping and Range of Ions in Matter, online available at: `https://www.srim.org/` (accessed December 08, 2017). 121

[100] Y. Kwon, Y. B. Yun, J. M. Ha, J. S. Lee, Y. S. Yoon, J. W. Eun, "Radiation damage of multipixel Geiger-mode avalanche photodiodes irradiated with low-energy γ's and electrons", Journal of the Korean Physical Society 60 (10) (2012) 1803–1808. `doi:10.3938/jkps.60.1803`. 123, 141

[101] H. Jafari, S. A. H. Feghhi, "Analytical modeling for gamma radiation damage on silicon photodiodes", Nuclear Instruments and Methods in Physics Research, Section A: Accelerators, Spectrometers, Detectors and Associated Equipment 816 (2016) 62–69. `doi:10.1016/j.nima.2016.01.079`. 123, 141

[102] I. Nakamura, "Radiation damage of pixelated photon detector by neutron irradiation", Nuclear Instruments and Methods in Physics Research, Section A: Accelerators, Spectrometers, Detectors and Associated Equipment 610 (1) (2009) 110–113. `doi:10.1016/j.nima.2009.05.086`. 141

[103] Y. Musienko, A. Heering, R. Ruchti, M. Wayne, A. Karneyeu, V. Postoev, "Radiation damage studies of silicon photomultipliers for the CMS HCAL phase I upgrade", Nuclear Instruments and Methods in Physics Research Section A: Accelerators, Spectrometers, Detectors and Associated Equipment 787 (Supplement C) (2015) 319 – 322, new Developments in Photodetection NDIP14. doi:10.1016/j.nima.2015.01.012. 141

[104] V. Bindi, A. D. Guerra, G. Levi, L. Quadrani, C. Sbarra, "Preliminary study of silicon photomultipliers for space missions", Nuclear Instruments and Methods in Physics Research Section A: Accelerators, Spectrometers, Detectors and Associated Equipment 572 (2) (2007) 662 – 667. doi:10.1016/j.nima.2006.12.011. 141

[105] P. F. Bloser, J. S. Legere, L. F. Jablonski, C. M. Bancroft, M. L. McConnell, J. M. Ryan, "Silicon photo-multiplier readouts for scintillator-based gamma-ray detectors in space", in: IEEE Nuclear Science Symposuim Medical Imaging Conference, 2010, pp. 357–360. doi:10.1109/NSSMIC.2010.5873780. 141

[106] Z. Li, Y. Xu, C. Liu, Y. Gu, F. Xie, Y. Li, H. Hu, X. Zhou, X. Lu, X. Li, S. Zhang, Z. Chang, J. Zhang, Z. Xu, Y. Zhang, J. Zhao, "Characterization of radiation damage caused by 23 MeV protons in Multi-Pixel Photon Counter (MPPC)", Nuclear Instruments and Methods in Physics Research, Section A: Accelerators, Spectrometers, Detectors and Associated Equipment 822 (2016) 63–70. doi:10.1016/j.nima.2016.03.092. 141

[107] E. Garutti, R. Klanner, D. Lomidze, J. Schwandt, M. Zvolsky, "Characterisation of highly radiation-damaged SiPMs using current measurements" arXiv:1709.05226. 141

[108] R. Hofstadter, J. A. McIntyre, "Measurement of Gamma-Ray Energies with Two Crystals in Coincidence", Phys. Rev. 78 (1950) 619–620. doi:10.1103/PhysRev.78.619. 141

[109] G. Nelson, D. Reilly, "Gamma-Ray Interactions with Matter", in: D. Reilly, N. Ensslin, H. Smith (Eds.), Passive Nondestructive Analysis of Nuclear Materials, Los Alamos Laboratory, pp. 27-42 (NUREG/CR-5550, LA-UR-90-732), 1991, Ch. 2. 141

[110] G. Lindström, "Radiation damage in silicon detectors", Nuclear Instruments and Methods in Physics Research Section A: Accelerators, Spectrometers, Detectors and

Associated Equipment 512 (1) (2003) 30 – 43, proceedings of the 9th European Symposium on Semiconductor Detectors: New Developments on Radiation Detectors. doi:10.1016/S0168-9002(03)01874-6. 142

[111] I. Pintilie, G. Lindstroem, A. Junkes, E. Fretwurst, "Radiation-induced point- and cluster-related defects with strong impact on damage properties of silicon detectors", Nuclear Instruments and Methods in Physics Research Section A: Accelerators, Spectrometers, Detectors and Associated Equipment 611 (1) (2009) 52 – 68. doi:10.1016/j.nima.2009.09.065. 142

[112] M. Moll, "Radiation Damage in Silicon Particle Detectors - Microscopic defects and macrosopic properties", Hamburg University, 1999. 148

[113] M. C. Vignali, E. Garutti, R. Klanner, D. Lomidze, J. Schwandt, "Neutron irradiation effect on SiPMs up to $\Phi_{neq} = 5 \cdot 10^{14}$ cm^{-2}", Nuclear Instruments and Methods in Physics Research Section A: Accelerators, Spectrometers, Detectors and Associated Equipmentdoi:10.1016/j.nima.2017.11.003. 148

List of Publications

First-author contributions

E. Engelmann, F. Wiest, D. Sporea, A. Stancalie, D. Negut, E. Garutti, W. Hansch, "Investigation of Radiation Hardness of SiPM Exploiting the Effect of Hot Carrier Luminescence", in: Talk given at International Conference on New Developments in Photodetection (NDIP), Tours, France, 2017

E. Engelmann, F. Wiest, P. Iskra, W. Hansch, A. Stancalie, D. Sporea, D. Negut, "Test of KETEK PM1150T SiPM under low-dose ^{60}Co gamma-irradiation", in: Talk given at COST Action TD1401 Annual meeting, Larnaca, Cyprus, 2017

E. Engelmann, E. Popova, S. Vinogradov, F. Wiest, E. Garutti, W. Hansch, "Impact of Local Defects on the Dark Count Rate of KETEK SiPM", in: Talk given at IEEE Nuclear Science Symposium, Strasbourg, France, 2016

E. Engelmann, S. Vinogradov, E. Popova, F. Wiest, P. Iskra, W. Gebauer, S. Loebner, T. Ganka, C. Dietzinger, R. Fojt, W. Hansch, "Extraction of activation energies from temperature dependence of dark currents of SiPM", Journal of Physics: Conference Series, 675 (4) (2016) 042049. doi:10.1088/1742-6596/675/4/042049

E. Engelmann, S. Vinogradov, E. Popova, F. Wiest, P. Iskra, T. Ganka, C. Dietzinger, W. Gebauer, S. Loebner, R. Fojt, W. Hansch, "Temperature Dependent Investigation of DCR of SiPM", in: Talk given at International conference on particle physics and astrophysics (ICPPA), Moscow, Russia, 2015

E. Engelmann, S. Vinogradov, E. Popova, F. Wiest, P. Iskra, T. Ganka, C. Dietzinger, W. Gebauer, S. Loebner, R. Fojt, W. Hansch, "Extraction of Activation Energies from Temperature Dependent Investigations of Dark Current", in: Talk given at International Conference on New Photo-Detectors (Photodet), Moscow, Russia, 2015

Co-author contributions

W. Gebauer, E. Engelmann, Th. Ganka, P. Iskra, S. Loebner, A. M. Seco, F. Wiest, R. Fojt, W. Hansch, "Evaluation of a reliable and cost-effective PDE measurement setup with short cycle time for SiPM production monitoring", Nuclear Instruments and Methods in Physics Research Section A: Accelerators, Spectrometers, Detectors and Associated Equipment (2017). doi:10.1016/j.nima.2017.10.062

A. Stancalie, D. Sporea, D. Ighigeanu, E. Engelmann, F. Wiest, P. Iskra, W. Hansch, "Investigation on electron beam radiation defects induced in KETEK PM3350 silicon photomultipliers", in: Talk given at International Conference on Scintillating Materials and their Application (SCINT), Chamonix, France, 2017

D. Sporea, E. Engelmann, T. Ganka, F. Wiest, "Testing facilities to support FAST COST Action", in: Talk given at FAST WG3 & WG4 meeting, Lisbon, Portugal, 2017

E. Popova, S. Ageev, D. Philippov, P. Buzhan, A. Stifutkin, S. Klemin, P. Iskra, W. Butler, E. Engelmann, F. Wiest, R. Fojt, F. Kayumov, "Active SiPM-Fast Analogue CMOS SiPM Prototypes with Integrated Amplifiers", in: Talk given at IEEE Nuclear Science Symposium, Strasbourg, France, 2016

A. Kolb, E. Engelmann, P. Major, G. Patay, B. Tölgyesi, C. Parl, T. Ganka, M. J. Czeller, G. Nemeth, B. Pichler, "Preclinical PET Detector with Temperature Gain Compensation", in: Talk given at IEEE Nuclear Science Symposium, Strasbourg, France, 2016

W. Hartinger, C. Dietzinger, T. Ganka, F. Schneider, P. Iskra, E. Engelmann, S. Loebner, W. Gebauer, A. M. Seco, F. Duesberg, F. Wiest, "Introduction of KETEK's latest SiPM generation Evaluation Kits", in: Talk given at IEEE Nuclear Science Symposium, Strasbourg, France, 2016

W. Hartinger, F. Wiest, J. Knobloch, R. Fojt, E. Popova, C. Dietzinger, T. Ganka, P. Iskra, E. Engelmann, S. Loebner, W. Gebauer, A. M. Seco, F. Duesberg, "Recent Enhancement of the KETEK SiPM Device Performance with regard to Timing, Crosstalk and CMOS Compatibility", in: Talk given at International conference on particle physics and astrophysics (ICPPA), Moscow, Russia, 2015

F. R. Schneider, T. R. Ganka, G. Seker, E. Engelmann, D. Renker, S. Paul, W. Hansch, S. I. Ziegler, "Characterization of blue sensitive 3x3 mm^2 SiPMs and their use in PET", Journal of Instrumentation 9 (7) (2014) P07027. doi:10.1088/1748-0221/9/07/P07027

F. Wiest, E. Engelmann, C. Dietzinger T. Ganka, W. Gebauer, W. Hartinger, P. Iskra, S. Loebner, A. Marquez-Seco, N. Miyakawa, "SiPM performance boost for scintillating fiber tracker upgrade at CERN LHCB", in: Talk given at IEEE 2014 Nuclear Science Symposium, Seattle, USA, 2014

Th. Ganka, Ch. Dietzinger, W. Gebauer, E. Engelmann, P. Iskra, F. Wiest, R. Fojt, W. Hansch, "Influence of Active Area and Different Read-Out Mechanisms on the Timing Performance of Silicon Photomultipliers", in: Talk given at IEEE 2014 Nuclear Science Symposium, Seattle, USA 2014

Acknowledgements

First of all I would like to thank Prof. Dr. Walter Hansch for giving me the opportunity to work on this interesting and challenging topic. I am thankful for his support and his open mind which enabled me to realize my own ideas throughout this project.

Sincere thanks to Prof. Dr. Erika Garutti for supervising this work. Our frequent communication helped me to keep track of the performed measurements and to put the obtained results in relation with each other. My visits to Hamburg University were always instructive and gave me the opportunity to get valuable insight into the application of SiPMs in physical research projects.

I would like to thank the management team of KETEK GmbH, Silvia Wallner, Dr. Reinhard Fojt and Dr. Jürgen Knobloch for enabling this PhD thesis. Their constant support paved the way for the successful outcome of this research project.

Great thanks to Dr. Florian Wiest and Dr. Peter Iskra. Without their shared knowledge and constant participation, I would not have been able to complete this work. They were always available for detailed and open minded discussions which created a stimulating working environment. I also appreciate that they were willing to try out the ideas which we generated during these discussions.

I owe a huge thanks to my mentors Dr. Elena Popova and Dr. Sergey Vinogradov. They supported me with their great expertise and ideas throughout this work. Their visits to Munich and my visits to Moscow were extremely important for me to improve as a scientist. I appreciate their dedication, patience and persistence which always motivated me to walk the "extra mile".

BIBLIOGRAPHY

Dr. Dieter Renker introduced me to the field of solid-state photodetectors during my Master's studies and inspired me to go in this direction, for which I am thankful. Our detailed discussions and the time at his lab were always a pleasure for me.

I thank Dr. Thomas Ganka who made me familiar with the SiPM-characterization and supervised my first LabVIEW-projects during my Master's studies. After he finalized his own PhD project, I could take over his well organized laboratory. Thanks for the great time we spent together in the lab and beyond.

The work of Dr. Torsten Sulima, who took care of the administrative tasks is greatly appreciated.

I also thank the PhD students Sabrina Löbner and Wolfgang Gebauer for the pleasant atmosphere in our office and wish both good luck with their theses.

My deepest gratitude goes to my family for their love and support. Especially, I thank my parents for enabling and encouraging me to study, for their guidance and their endless patience throughout the years. Thanks to my brother, Artur, for his unfailing emotional support and for being in tune with me all my life. He helped me in whichever way he could and is in many aspects my role model. Last, but by no means least, I thank my wife, Nina, for her love, support and understanding. She was never tired of listening to me talking about my experiments for hours and hours. I will forever be thankful to her for giving me the greatest gift in my life, our son Emil.

www.ingramcontent.com/pod-product-compliance
Lightning Source LLC
Chambersburg PA
CBHW070719220326
41598CB00024BA/3228